Hybrid Modernity

This book provides a detailed historical and design analysis of the development of parks and modern landscape architecture in late 20th century China. It questions whether the fusion of international influences with the local Chinese design vocabulary in late 20th century China has created a distinctive and novel approach to the design of public parks.

Hybrid Modernity proposes a new theory for examining the design of public parks built in post-Mao China since the reforms and sets the various processes for China's late 20th century socio-cultural context. Drawing on modernization theory, research on China's modernity, local and global cultural trends, it illustrates through a range of case studies ways hybrid modernity defines a new design genre and language for the spatial forms of parks that emerged in China's secondary cities. Featured case studies include the Living Water Park in Chengdu, Sichuan Province, Zhongshan Shipyard Park in Guangdong Province, Jinji Lake Landscape Master Plan in Suzhou, Jiangsu Province, and the West Lake Southern Scenic Area Master Plan in Hangzhou, Zhejiang Province. This book argues that these forms represent a new stage in China's history of landscape architecture. The work reveals that as a new profession, landscape architecture has greatly contributed to China's massive urban experiment.

This book is an ideal read for students enrolled in landscape architecture, architecture, fine arts and urban planning programs who are engaged in learning the arts and international design education.

Mary G. Padua, PhD, RLA, is an internationally recognized educator, thought leader, contemporary theorist and artist. Her research focuses on socio-cultural phenomena, human-centered design and the meaning of public space. She is one of the first English language writers on post-Mao designed landscapes and the discipline of landscape architecture as contributing agents to China's late 20th century urban experiment. She maintains MGP Studio, a critically minded practice rooted in craft, equity, inclusion and restorative experiences, and teaches at Clemson University's School of Architecture.

Studies in Architecture Series

Series Editor: Eamonn Canniffe, Manchester School of Architecture, *Manchester Metropolitan University, UK*

The discipline of Architecture is undergoing subtle transformation as design awareness permeates our visually dominated culture. Technological change, the search for sustainability and debates around the value of place and meaning of the architectural gesture are aspects which will affect the cities we inhabit. This series seeks to address such topics, both theoretically and in practice, through the publication of high quality original research, written and visual.

Other titles in this series

On Surface and Place
Between Architecture, Textiles and Photography
Peter Carlin

Architecture, Death and Nationhood
Monumental Cemeteries of Nineteenth-Century Italy
Hannah Malone

Reconstruction and the Synthesis of the Arts in France, 1944–1962
Nicola Pezolet

Irish Housing Design 1950–1980
Out of the Ordinary
Brian Ward, Michael Pike, Gary Boyd

Hybrid Modernity
The Public Park in Late 20th Century China
Mary G. Padua

Le Corbusier's Practical Aesthetic of the City
The Treatise 'La Construction des villes' of 1910/11
Christoph Schnoor

For more information about this series, please visit: https://www.routledge.com/architecture/series/ASHSER-1324

Hybrid Modernity

The Public Park in Late 20th Century China

Mary G. Padua

LONDON AND NEW YORK

First published 2021
by Routledge
2 Park Square, Milton Park, Abingdon, Oxon OX14 4RN

and by Routledge
52 Vanderbilt Avenue, New York, NY 10017

Routledge is an imprint of the Taylor & Francis Group, an informa business

© 2021 Mary G. Padua

British Library Cataloguing-in-Publication Data
A catalogue record for this book is available from the British Library

Library of Congress Cataloging-in-Publication Data
Names: Padua, Mary, author.
Title: Hybrid modernity: late 20th century parks in China / Mary G. Padua.
Description: Abingdon, Oxon; New York, NY: Routledge, 2020. |
Series: Ashgate studies in architecture | Includes bibliographical references and index.
Identifiers: LCCN 2020011047 (print) | LCCN 2020011048 (ebook) |
ISBN 9781472445674 (hardback) | ISBN 9781315587684 (ebook)
Subjects: LCSH: Parks–China–Design–History–20th century. |
Landscape architecture–China–History–20th century.
Classification: LCC SB484.C55 P33 2020 (print) | LCC SB484.C55 (ebook) |
DDC 712.50951–dc23
LC record available at https://lccn.loc.gov/2020011047
LC ebook record available at https://lccn.loc.gov/2020011048

ISBN: 9781472445674 (hbk)
ISBN: 9781315587684 (ebk)

Typeset in Sabon
by Deanta Global Publishing Services, Chennai, India

Dedicated to my students and those interested in the transformative power of cultural (designed, vernacular, agrarian and experimental) landscapes within the public gaze

Contents

Figures and tables

Figures

Tables

Preface

The making of this book involved a long gestation period with support from numerous individuals and institutions. It evolved from early writings on award-winning parks in China, a rigorous scholarly journey involving the granting of the PhD from the Edinburgh College of Art, as well as a range of discussions with designers and educators in China, and my network of artists, poets and scholars from my time at the University of Hong Kong (HKU). Numerous discussions at various academic gatherings around the world brought to light the vitality of advancing theory-building and praxis-based scholarship in the discipline of landscape architecture and broader area of built environment studies, as well as the benefits of bridging the scholarly literature between China and the West.

As the first English language book on the evolution of public parks in modern China with a focus on the late 20th century, it traverses beyond the traditional history of park design with a "transdisciplinary gaze" and synthetic theory deemed "hybrid modernity". In this context, the public park is characterized as part of post-Mao cultural production and "growth first" period of hybrid modernization and the rise of the secondary city in China. The synthesis of hybrid modernity draws from the convergence of international influences, the Chinese Picturesque garden tradition and literati culture, modernization, nation-building and the search for identity after China opened to the postmodern world. Case study analysis of four purpose-built public parks serves as the schema for this new design language. It weaves an interpretative narrative for post-Mao China's tremendous urban experiment and orients the reader to China as a world civilization with its own distinctive trajectory.

In terms of support and book production, key acknowledgments include my research assistants: Szeto Chi Wah, Yao Jiachung, Liu Xiaotong and Wu Jueminsi. Dr. Bao Zhiyi from Zhejiang Agriculture and Forestry University, Yu Kongjian from Peking University and Turenscape, and Betsy Damon were instrumental in creating the necessary introductions, especially the late Sun Xiaoxiang (1921–2018) from Beijing Forestry University. I am grateful to staff at the Pei Partnership who arranged "tea" and a meeting with I. M. Pei (1917–2019); and the many individuals engaged in the design of the four case study parks who took the time to discuss their experiences and share visual materials for use in this book. This includes Yu and Damon, mentioned earlier, Sean Chiao and David Jung, currently in leadership positions at AECOM, and the Hangzhou Landscape Architecture Design Institute. HKU provided research grants and study leave with the support of Tom Kvan, then Dean of HKU's Faculty of Architecture who encouraged me to pursue the PhD. Catharine Ward Thompson was instrumental during my studies at Edinburgh College of Art. The University of

Florida provided support and time for me to spend at Dumbarton Oaks. Clemson University also provided support. I am indebted to Jim Robins, my life-long partner, for his emotional support and thoughts on social theory. I am forever grateful to my late father who nurtured my hunger for knowledge and my mother for her unconditional love. Lastly, I am grateful to the Routledge team for their patience, especially Grace Harrison.

1 Introducing hybrid modernity

The tremendous post-Mao late 20th century nation-building effort in the People's Republic of China (China) has been and continues to be a complex topic of study among cultural theorists, art critics and historians, urban scholars and social scientists operating in both China and western nations. The post-Mao period commenced with the 1976 demise of Chairman Mao Zedong, China's highly revered revolutionary communist leader, and subsequent rise of Deng Xiaoping, paramount leader and father of post-Mao revolutionary reforms. Also known as "Reform and Opening-up, *Gaige Kaifang*", Deng's late 1970s open door policy evolved from the "Four Modernizations, *Sige Xiandaihua*", a nation-building policy that targeted four sectors: agriculture, industry, national defense, and science and technology. Deng shifted China from Mao's hardline socialist approach to a market economy open to foreign investment. The ensuing post-Mao China experience involved a vast transformation with the fastest and most extensive urbanization in world history. As the architectural critic Ouroussoff (2008) argued, historical cycles that took a century or more to unfold in Europe or North America can be compressed into less than a decade in China.

While scholarly work has examined aspects of China's late 20th century built environment, limited work has examined the design history and theoretical evolution of post-Mao public parks, especially their spatial forms. This book intends to fill this knowledge gap through an investigation of the public park, one of two archetypal spatial forms within the discipline of landscape architecture – the private garden is the companion archetype. It builds the multivalent theory, "hybrid modernity" and related socio-cultural theory, "hybrid modernization" through the case study analysis of four late 20th century purpose-built parks. This theoretical synthesis suggests these parks emerged as part of China's tremendous urban experiment and symbolized notions of local identity and nation-building through the fusion of international design influences and the Chinese Picturesque garden tradition. It also suggests broader transformations in the post-Mao period caused a competitive milieu for local government leaders and the rise of the secondary city – a vehicle for entrepreneurial activities and design innovation.

The public park, generally, is a designated tract of land and open space in a city, usually a vegetated and designed outdoor environment, set aside for the general public's recreational and leisure activities. Within this urban context, a public park is intended for non-productive agricultural purposes. Purpose-built public parks historically emerged as part of 19th century modernization and city-making practices in Europe and North America, and especially as a public health response to industrialization.

In the same period, royal gardens on private land for royal use only, located in or around rapidly growing western cities, were transformed and adapted for public use. Public parks also fall under the broader umbrella of public space – accessible outdoor open space where people can gather for leisure and social activities. This category of public space typically includes parks, squares, plazas, commons, markets, waterfront promenades and the connecting spaces, streets, boulevards and related pedestrian paths or sidewalks. While the design and study of public parks fall largely within the domain of landscape architecture praxis and research, scholars and professionals operating within the realm of city planning and urban design consider the public park an important community health component, as well as fundamental to the physical form and socio-cultural fabric of the modern city. It is widely accepted that scholarly investigations of the public park fall neatly within the broader literature on the built environment and urbanism.

Since formally starting this research in 2005, the English-language literature on China's public park evolution, generally, or how it contributed to the built form of cities in the post-Mao 20th century period has been limited. This is still the case nearly fifteen years later. Dong (1999) and Esherick (1999) discuss the introduction of public parks in Republican China as part of their transformation of imperial urban spaces and modern city-making practices circa 1912–1916. Cranz (1979) reflects on the function and consumption of parks in socialist China as conveyed to her by government officials during her visit there. Friedmann's (2005) book presented the first deep investigation into China's late 20th century process of urbanization and its multi-layered meaning with public space only mentioned in a footnote. Yu and Padua's (2006) edited book introduces concepts and challenges facing landscape architecture in China at the turn of the new millennium, and projects by Turenscape, the Beijing-based private design and planning firm established by Yu Kongjian, a native-born Chinese educated at Harvard University who has developed an international reputation. Urban fever and China's post-Mao cosmetic cities are framed in terms of the production of superficial, mimetic and ornamental neo-classical parks (Yu & Padua 2007). Padua (2007) synthesizes a post-traditional design identity for designed outdoor public spaces in Hong Kong's designated redevelopment projects. Campanella's (2008) expansive book on China's urbanization discusses theme parks and landscapes of consumption in suburban post-Mao China. Lu (2006) examines China's urban geography and built environment in the period spanning 1949–2005 and weaves a discursive narrative around scarcity and modernity within a socialist space. Rinaldi (2011) discusses the influence of traditional Chinese gardens on China's contemporary landscape architecture. Wu and Gaubatz (2013) provide an exhaustive account of the evolution of the Chinese city and touched on the function of public parks. Padua (2003, 2004a, 2004b, 2006, 2008, 2012, 2013a, 2013b, 2014) reviewed award-winning designed landscapes completed during China's late 20th century and the first decade of the 21st century.

Where the aforementioned publications offer an understanding of Mao and post-Mao urban contexts, this book examines and interrogates more closely the design, spatial forms and meaning of urban public parks in late 20th century mainland China. It also introduces the first English language account of modern China's history of the public park, especially as part of city-making and cultural production. It reveals the awakening of modern landscape architecture praxis as significant actors for post-Mao cultural production and the making of late 20th century China's built environment. It brings to light park design innovations in China's secondary cities by the late 1990s.

The book acknowledges that China's primary cities, Beijing, Shanghai and Guangzhou (formerly known as Canton), have their own colonial heritage of park-making for foreign consumption and international reputations when China was a world player in international trade and cultural production (textiles, porcelain, tea, 18th century *Chinoiserie*, art and antiquities, and other commodities). Shanghai's urban heritage is recognized as an international and cosmopolitan place of modernity in the years before the 1937 Japanese occupation for both Chinese and foreign expatriates. China's late 20th century opening to the world gave rise to its secondary cities as vessels for innovation and experimentation in the making of purpose-built public parks.

The author's early research on public parks in post-Mao urban China was based on the premise that landscape architecture praxis is a design discipline. Hence, it was conceptualized as an exercise in design inquiry and history – interpreting the designed landscape or outdoor designed environment as a cultural product formed by the social, economic and political circumstances of a particular society or period. This follows the intellectual tradition associated with the author's former professors, J.B. Jackson (1970), Charles Jencks (1978), Dolores Hayden (1995) and Catharine Ward Thompson (1998). Following this trajectory, studying parks as cultural by-products of post-Mao China necessitated the formulation of a theoretical and analytical framework for understanding the social, economic and political situation of that period. However, early on it became clear this complex post-Mao multi-dimensional construct for examining urban parks was not elucidated for the landscape architecture audience. Furthermore, the author discovered the public park's evolution in the Mao or pre-Mao periods and complexity of modern China's 19th and 20th centuries had not been made available to this same English-speaking audience. This book offers a transdisciplinary and interpretative narrative for the ideological context and evolution of public parks in modern China. It introduces the Chinese Picturesque design genre, especially as an element of cultural identity, and discusses international influences from the late 20th century western world. In this context, the synthesis of the theory, hybrid modernity, posits purpose-built parks as intertwined with the fusion of local and international design influences, modern nation-building, and the search for local cultural identity during China's post-Mao late 20th century period.

With the limited English language literature on the design history of China's late 20th century public parks, the author drew from scholarly works on modernization theory, and China's modernity as a multivalent phenomenon embedded with socio-cultural, political and economic dimensions. This transdisciplinary approach looked "inside" China or scholarly work discussing major themes during its late 20th century cultural development, as well as the broader international English language literature on cultural trends "outside" China or the postmodern West. The convergence of ideas espoused by social scientists, cultural theorists and art and design historians investigating late 20th century China, and broader international cultural trends, inform the narrative to characterize China's late 20th century period.

The concept of modernization in itself originally emerged out of the domain of sociology and has been attributed to the iconic social theorist, Max Weber, and one of his followers, Jürgen Habermas (White 1989). Their intellectual roots stemmed from the Frankfurt School, a social research institute that existed in pre-World War II Germany. However, social theory and the social sciences, like all scholarly fields, evolve over time. Hence, aspects of research in this book expand from the Frankfurt

School's earlier thinking on modernization theory, especially later work espoused by theorists operating within the realm of post-colonial cultural studies and the globalization discourse (Appadurai 1990; Bhabba 1994; Dirlik 2002).

Appadurai (1990), Bhabba (1994) and others argue that social and economic change can be simultaneously influenced by global forces and remain culturally idiosyncratic. In their perspective, the process of change is embedded within culture; but the societies in which change is taking place are not isolated from the larger world. In the opposing position, globalization can be a major force for change without compromising the uniqueness of local cultures. The results of change are not determined solely by global forces, they also reflect local culture. The process of modernization and the form that modernity takes thus may be specific to individual societies. This book builds from this premise, particularly China as a world civilization with its own unique societal evolution.

Concepts of nation-building, consumerism and identity, experienced among the cultural intelligentsia in China during the 1980s and 1990s, are elucidated in the available literature on post-Mao cultural development. This book explores these concepts and themes as a way to orient the reader to post-Mao late 20th century China; and these also inform the foundation for hybrid modernity and hybrid modernization. In the normative language of social science, cultural studies, humanities and design research, epistemologically this study is qualitative, interpretative and transdisciplinary. The goal is to cast light on the meaning of purpose-built public parks in China's late 20th century through a theoretical framework that represents the convergence of scholarly thought beyond the discipline-specific perspectives.

Some research questions below guided earlier thinking on late 20th century parks in urban China. Are the spatial forms represented in these parks new? Do they reflect larger social and cultural transformations that have taken place since China opened to the world? How and when did public parks evolve in China and what did they mean? In what ways do they reflect the profession of landscape architecture and education for landscape architects in China? The answers to these questions provided a means for addressing the larger question that lies at the heart of this book: has the fusion of international influences with a local Chinese design tradition created a distinctive approach to the design of public parks that is novel in the post-Mao context? How has this taken place, and what does it mean for landscape architecture during China's late 20th century period of hyper-rapid urbanization?

The process of hybrid modernization creates a crucible for answering these questions about late 20th century urban China, and its synthesis will be discussed more deeply and explored in the book. However, as a matter of introduction and situating the reader, hybrid modernity and the process of hybrid modernization are posited as a temporal interpretation of China's 1976 to 2001 period – bracketed on one side with Mao's demise and Deng Xiaoping's revolutionary reforms, and 2001 on the other side when the International Olympic Committee announced Beijing as the 2008 Olympic host city, and more significantly the World Trade Organization's (WTO) action to formalize China's membership. This latter temporal bookend combines the beginning of the new millennium, selection of Beijing for the 2008 Olympics, and complex political and economic moves required for WTO membership. Furthermore, it strongly suggests the commencement of another period of study.

China's late 20th century period of accelerated or hyper-rapid urbanization can be demonstrated by the number of cities: 1976 = 190+ cities; 1984 = 290+ cities;

2001 = 560+ cities; it reflects approximate urbanization rates in 1990 at around 26%, with nearly 36% in 2006 and over 56% by the end of 2016 (Padua 2014; Wu et al. 2016). These descriptive statistics infer the rise of the secondary city at the end of China's 20th century. The author argues that these secondary cities or non-primary cities emerged as places for entrepreneurship and design innovation. It is within these secondary cities that mayors advocated for the design of purpose-built public parks as economic catalysts for foreign investment and marketing their cities as landmarks for political and cultural purposes. It's important to acknowledge economic analysts and social scientists discuss a tiered system of cities in China based on gross national product. However, given the widely accepted notion of China's unreliable data, the author argues that secondary cities emerged as living laboratories where purpose-built parks were part of post-Mao China's vast urban experiment.

The concept of hybrid modernity is partially situated within the realm of post-colonial and cultural studies. It particularly expands from Appadurai's (1990) theory on alternative modernity which positions China in the worldview as a long-standing civilization and nation that is socio-culturally distinctive. Along this scholarly trajectory, the broader development of China in the late 20th century is interpreted and analyzed on its own and not within the dichotomous narrative for nations, e.g. advanced/advancing, first world/third world, developed/developing, and so forth. This seemingly pejorative binary discourse evolved in western-based thinking within the field of development and modernization studies, and has no utility in building the argument and conceptualizing China's late 20th century hybrid modernity.

In terms of examining Chinese identity, Anderson (1991) notes it is essentially a mythological construction, largely serving political purposes. Nationalism or nation-building – "Chineseness" as local identity – is inter-related with Anderson's notion of a re-imagined post-Mao national identity. On the limitations of conceptualizing Chinese identity, it's important to note the social norm for ethnic identity is extremely complex and depending on geographic, economic and socio-cultural circumstances highly contested within China. Over fifty so-called ethnic minority groups live throughout China with the majority population dominated by the Han ethnic group. For this exploration, the dominance of both Han influence and Confucian ideology contribute to the broader definition of Chinese identity. Confucianism is attributed to China's literati and imperial court culture with government scholar-officials required to pass a series of so-called civil service examinations based on the Confucian classics. Mao attempted to aggressively "erase" this culture and in the earlier late Qing/Republican era, attempts were made to modernize the feudal imperial government by retaining the essence of Chineseness while learning from the west – a highly complex ongoing ideology and recurring theme for modern China.

When China opened to the world in the late 20th century and charted Deng's New Era, *Xin Shiqi*, the western cultural narrative and discourse were postmodern. Cultural theorists and art historians have carefully debated the question of China's position in the western postmodern discourse, especially in light of the broader debate on China's modernity and 20th century revolutionary praxis (Zhang 1997; Dirlik 2002). These temporal and ideological debates are extremely complex and will be touched on in the book. However, in keeping with this transdisciplinary and interpretative approach for investigating the post-Mao late 20th century era, the concept for hybrid modernity suggests an alternative to the western postmodern context. It temporally follows Mao's utopian version of modernity, a socialist period of collective identity, cultural propaganda and nation-building, chaos and scarcity – an overall period of extremes and contradictions (Liu 1971).

This book discusses new spatial forms of purpose-built parks in China's late 20th century as a major evolutionary shift forward from the superficiality represented in Yu and Padua's (2007) outdoor designed environments. These public parks represent a new hybrid modern design language – the interplay of global and local design influences along with nation-building, modernization and notions of Chinese identity. Post-Mao landscape architecture practice by Chinese local design institutes was largely influenced by the traditional Chinese garden – private residential gardens designed and built by retired scholar-officials from China's imperial court. The author refers to this design genre, language and imagery as the "Chinese Picturesque". After China opened to the world, international design trends were introduced to contemporary landscape architecture practice by foreign-based professionals; students who returned from studying abroad; and government and Communist Party of China (CPC) officials who traveled abroad and returned with photographs of projects and places they visited. Furthermore, China was bombarded with a variety of western media and publications that were quickly translated and disseminated via analog and digital means.

In the late 20th century period, geographically in the west, praxis, theory-building and research within the discipline of landscape architecture were wide-ranging. Lawrence Halprin, a San Francisco-based landscape architect, and his award-winning practice pioneered engaging with communities through participatory workshops. His office employed over 100 employees at one point and his environmental design projects revitalized cities up and down the American Pacific Coast, Midwest and South, as well as around the world. Ian McHarg's development-oriented land planning and regional approach influenced a generation and expanded the field to consider natural resources and other factors in larger physical areas surrounding urban contexts through his map overlay approach and suitability method, the precursor to the Geographic Information Systems (GIS) computer software. The rise of the American corporate design firm occurred in parallel to projects pursued in the professional offices of Halprin and McHarg with award-winning projects by Sasaki & Associates, the SWA Group, EDAW and others who had branch offices around the nation. America's post-World War II era involved a population boom, a need for new residences and the subsequent rise of suburban development. In this light, the rise of corporate America occurred with the demand for "second homes", and some were located in purpose-built residential communities built around a golf course. In public park design innovation, the architect Bernard Tschumi and his submission for the international design competition for *Parc de la Villette* in Paris, France, created a stir among design critics, scholars and historians in the discipline of landscape architecture in the late 1980s. Questions were raised about professional territoriality, the credibility of an architect entering the realm of park design and a dialectic discourse emerged around the meaning of his deconstructivist and postmodern approach (Meyer 1991; Baljon 1999).

In response to the rise of corporate design, suburban development and the notion of American landscape architecture praxis as primarily "high" design similar to "high art" and its orientation to the commercial corporate realm, the concept of landscape urbanism (Connolly 1995; Waldheim 2006) emerged in the 1990s. In this re-situating of landscape architecture praxis, Waldheim (2006) suggests outdoor landscapes, or open spaces – voids between the buildings in cities open to the sky – could act as the city's primary form-drivers. Subsequently, landscape urbanism became interlinked with landscape infrastructure, stormwater management and related performative qualities of outdoor designed environments within the public realm, and may have informed the 21st century ecological urbanism discourse.

Other late 20th century topical areas in landscape architecture praxis and research involved cultural resources, heritage preservation and the resurgence of memorials; deployment of the narrative and contemporary representation; place-making, meaning and the human experience; natural conservation, greenways and ecological corridors, and habitat restoration; brownfields reclamation, post-industrial landscapes and adaptive re-use; environmental justice and social equity; the utilization and development of software like G.I.S. (geographic information system) and digital-mapping as it applied to ecological and sustainable design and planning. Empirical studies in the area of human behavior, community development and the environment also emerged. The public park or civic open space, in general terms, was either the focus or backdrop for these areas of inquiry in western landscape architecture praxis and research. These works were developed for and made accessible to an English-speaking audience located in Australia, North America, the United Kingdom and Western Europe. The evolution of spatial forms, functions, user needs, aesthetics and management of parks throughout the west, as well as the various types of parks, have been analyzed, documented and published (Meyer 1991; Cranz 1991; Baljon 1999; Schenker 1995; Thompson 1998; Tate 2001; Bull 2006; Czerniak & Hargreaves 2007). This book intends to complement this literature and introduce the public park in modern China, generally, and in particular the post-Mao period of hybrid modernization.

In addition to introducing the reader to theories on modernization, modernity and modernism, and the public park's relationship to China's modernity and cultural production, the synthesis of hybrid modernity draws from interviews of Chinese educators and professionals in landscape architecture, allied professionals and Chinese officials; archival research; ethnographic methods involving field observations; and the case study analysis of four purpose-built public parks in post-Mao secondary cities. These parks represent the corpus of the study and schema for hybrid modernity. As noted earlier, the book in part draws from within or "inside" post-Mao urbanism and post-socialist society, and the history of modern China (Dirlik 2002; Zhang 1997; Wang 2003; Friedmann 2005; Gao 1998, 2011).

Post-Mao transformations: framing hybrid modernity

This book suggests the new design paradigm, hybrid modernity, particularly emerged in the making of purpose-built public parks during China's late 20th century. Charles Jencks (2002) has argued that major shifts in thinking in fields such as architecture are created by broader changes in culture and worldview in areas such as religion, politics, and science. Five key transformations of this type were fundamental to the social, cultural and political transformation of China in the post-Mao era and briefly discussed in the next section. The convergence of these shifts is critical to understanding the post-Mao period of hybrid modernization, China's urban experiment, especially the rise of secondary cities. The book's author argues that local government officials in post-Mao 1990s secondary cities were critical agents who enabled innovation and experimentation in the making of purpose-built public parks. Due to historical development, colonial park-making heritage, and politics, design experimentation in post-Mao primary cities, Shanghai, Beijing and Guangzhou, was less focused on public parks. Satellite towns, new purpose-built suburbs, and the amalgamation of rural areas were typical post-Mao urban expansion and planning practices undertaken around China's primary cities (Lu 2006; Yu & Padua 2007; Campanella 2008).

In orienting the reader to China's post-Mao period, it's important to note one of the most basic transformations in late 20th century China was its opening to the world after several decades of isolation. This decades-long period of social and economic isolation can be defined by several facts. The earlier Republic of China (ROC) established in 1912 after the 1911 Revolution led by Sun Yat-sen was short-lived. Governing and transforming a feudal society into a modern nation were difficult tasks, especially with periods of internal strife, including the Taiping and Boxer Rebellions and conflicts among warlords. Non-Chinese or western "barbarians" were not allowed to communicate directly with officials of the Qing imperial court and Chinese individuals who served as go-betweens or middlemen gained tremendous economic and political power – subsequently emerging as warlords. The Japanese conflict and occupation circa 1937–1945, and World War II contributed to China's decades of 20th century disunity and ROC failure; this period allowed time for Mao Zedong and his followers to organize the rural peasantry for a military and political revolution. Mao's effectiveness at mobilizing rural villagers against civil strife enabled him to establish the People's Republic of China and single-party (Communist Party of China – CPC) government by the last quarter of 1949. As the world witnessed, modernizing a rural society proved difficult for Mao politically, especially with his failed economic policies, famine and tremendous loss of life, and the Cultural Revolution (CR) when schools were closed from 1966–1976.

Numerous cultural artifacts were violently destroyed during the CR along with the public humiliation of individuals politically deemed *bourgeois* or part of the urban elite. Traditional Chinese gardens and architecture were ravaged. However, some artifacts were preserved by enlightened individuals and officials. For example, while conducting field research at West Lake, Hangzhou in the Zhejiang Province, the author's host pointed out Ming dynasty stone bridges and other physical artifacts in newly designed landscapes and noted these were previously dismantled, stored and hidden from the Red Guards during the CR. While Mao's CR is another vast scholarly topic on its own and given its impact on China's isolation to the world, some government-sanctioned activities are noted later in the book. One objective is to broaden the reader's understanding of the fundamental transformation from Mao's dark period of poverty, chaos, anarchy, extremes and contradictions, and world isolation to Deng's reforms – opening China to the world.

The second major shift involved China's decentralization from tight central government control and its massive bureaucratic apparatus to greater local government autonomy. In other words, central government initiatives and programs devolved fiscal responsibility to local governments like the Township and Village Enterprises – recognized by economists and urban scholars with the acronym TVE (Byrd & Lin 1990). Local governments and businesses thrived from 1978 through the 1990s as a result of these policies. TVEs were intended to replace the collectives or communal factories, also known as state-owned enterprises (SOE), often located in the periphery of urban areas. However, it is widely accepted that China's post-Mao period involved parallel and overlapping activities. For example, communal factories were present throughout the 1990s as TVEs flourished. China also established Special Economic Zones (SEZs), geographic locations designated for export-oriented industries and Foreign Direct Investment (FDI) in the 1980s. In subsequent years, new locations throughout China were officially opened for FDIs, e.g. various coastal cities, major river delta regions, as well as the establishment of economic and technological development zones. These areas spurred entrepreneurial and robust competitive activities among provincial leaders, mayors, and CPC party officials. This shift from central government to local autonomy contributed to China's hyper-rapid urbanization, soaring economic growth and the rise of secondary cities.

A third critical transformation was the shift from socialism to a form of market economy, known as "socialism with Chinese characteristics, *zhongguo tese shehui zhuyi*" – essentially opening China to the world economically, especially for foreign investment. This infusion of capitalist economic ideology and ethos created a far-reaching transformation in the values and principles of the Chinese people, particularly the younger generations. Scholars from several disciplines have discussed China's economic shift from socialism to a market economy as highly contested in both positive and negative terms. Generally, social and cultural theorists see this shift as problematic to rural society as the population is subsumed into urbanization (Dirlik 2002; Wang 2003). Foreign designers and planners embraced the positive impact of China's market economy with "urban fever" – a nation open to experimentation (Koolhaas 2002; Yu & Padua 2007; Campanella 2008). During China's 1980s economic boom, political elitism, graft and corruption ran rampant among various levels of government and CPC officials. This caused public dissatisfaction with student-led demonstrations for a national democracy movement. Eventually, China's response to this New Era "open society" took a notorious turn and led to the 1989 Tiananmen Square massacre in Beijing.

These movements toward local autonomy and capitalism contributed to another socio-cultural transformation that reached deep into Chinese society – the shift of the workplace from the collective or communal factory to the emphasis on the individual as a producer, entrepreneur and consumer. Decentralization and emphasis on the individual encouraged competition and local entrepreneurism, and contributed to the rise of secondary cities in China. This shift ushered in a new era of consumption and helped give legitimacy to the pursuit of individual wealth. China's post-Mao society was emboldened by one of Deng's famous quotes, "Poverty is not socialism. To be rich is to be glorious" (Davies & Raskovic 2017).

Finally, the most fundamental social and economic change in China was the transformation of the country from Mao's era when the nation experienced widespread poverty and frequent famine to a society where few individuals suffered absolute deprivation and many could afford leisure time. This can be seen as China's larger transformation – from a basic rural peasantry to an urban society with growing wealth. Explosive growth in productivity and the modernization of the workplace helped to create a six-day and later five-day work week for people who had never enjoyed a day without labor. This resulted in new leisure time and a dramatic increase in demand for recreational activities (Davis 1995). Local tourism increased, as well as a pressing need for public parks.

In summary, the book's premise for China's late 20th century hybrid modernity and social process of hybrid modernization is based on key economic, socio-cultural and political shifts (Figure 1.1). Deng's reforms spurred these transformations and eventually led to hyper-rapid urbanization, and especially the rise of the secondary city in China. One other important factor that cannot be ignored in post-Mao China is the

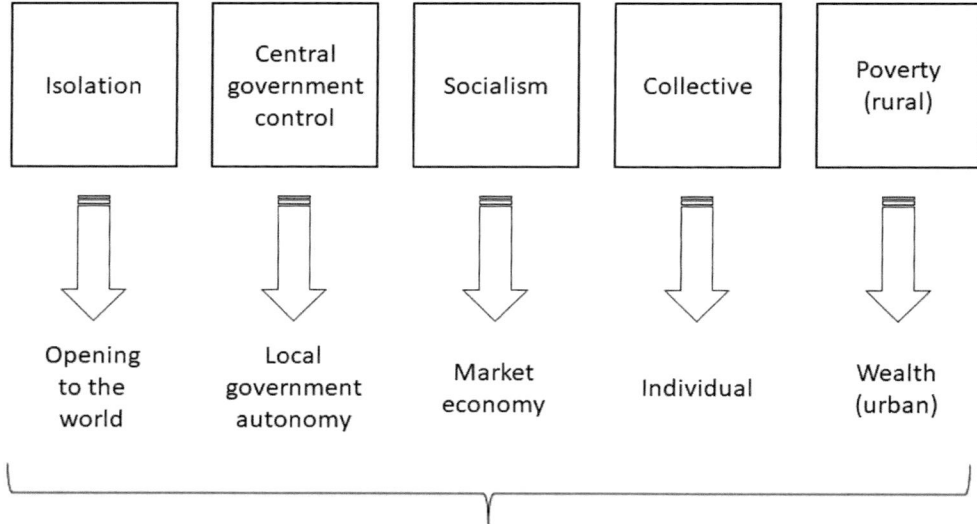

Late 20th century transformations in China - rise of the secondary city
Hybrid Modernization

Figure 1.1 Post-Mao China's transformations and late 20th century context, graphic by author

allure of politics and the agency of a single-party government. Ambitions among local government officials in China's late 20th century were tremendous and their eagerness to climb the political ladder is evident today. The creation of new purpose-built public parks was seen by local government leaders as a way to market their cities for foreign investment, build their legacy and advance political careers – tantamount to post-Mao nation-building. Furthermore, with China's propensity for rapid change, documenting these parks within this period was deemed critical, given the potential for their erasure.

China's physical setting and settlement patterns

China's vast physical landscape is defined by a variety of mountain ranges, plateaus, fertile valleys, and alluvial plains drained by a network of rivers that flow into its long eastern coastline. China's first cities were located in fertile basins and river valleys. Subsequently, the influence of foreign trade spurred urbanization and colonial development concentrated in commercial ports along the coast (Figure 1.2). China's spatial pattern of human settlement is a result of thousands of years of political territorialism, urbanization and periods of disunity, war and peace. Although late 20th century China's hybrid modernization process and accelerated urbanization involved the rise of new and secondary cities in compressed periods of time, it represents the continuation of a process that has ancient origins.

Since Deng's reforms, development policies shifted and evolved spatially across China. In broad terms, China attempted to mitigate the nation's so-called uneven development pattern. Initially, the accepted approach was to organize the nation spatially into three economic belts or regions, *sanda jingji didai*: eastern, central and western (Figure 1.3). Each region was assigned a role, respectively from east to west:

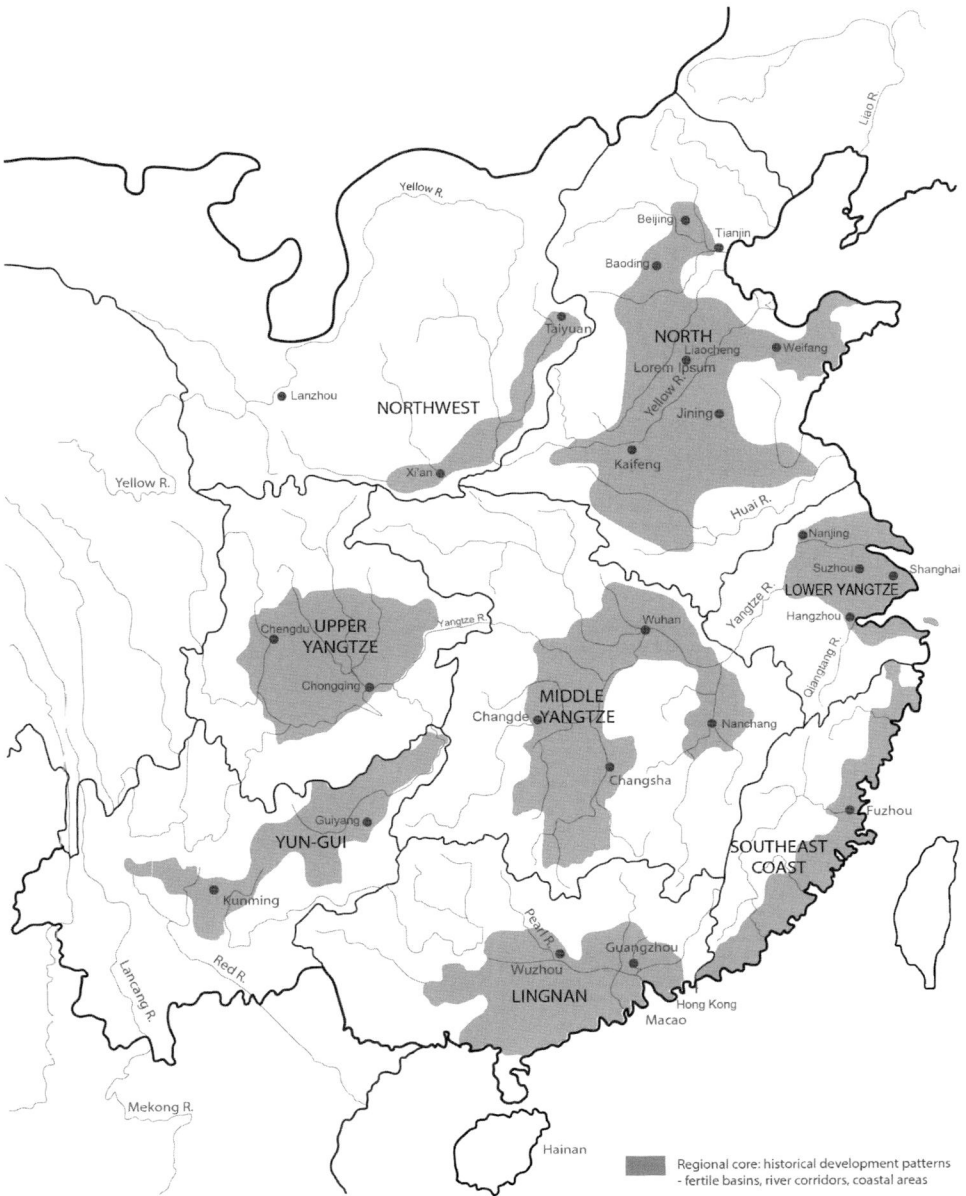

Figure 1.2 Generalized map of China's historical settlement patterns based on Skinner (1977), Ma and Wu (2005), and Padua (2014), graphic by X. Liu

export-oriented production and foreign trade; agriculture and energy production; and animal husbandry and mineral exploitation in the west; and regional sectoral and economic differences were referred to as "ladder-steps, *tidu*" (Fan 1995). The eastern region was also designated for high-technology and information-intensive industries. Other development policies were introduced through the 1990s. For example, with

Figure 1.3 Post-Mao China economic regions and open zones, partially based on Fan (1995) and Crane et al. (2018), graphic by X. Liu

the established five Special Economic Zones (SEZs) came designated "open coastal cities", open economic zones including the Pearl River and Yangtze River deltas, and other open cities and counties – all became spatial components of the "golden coastline, *huangjin*" (Fan 1995). China's 1980s and 1990s hyper-rapid urbanization and "growth first" policy brought massive environmental degradation and transformation of natural ecosystems throughout the nation. Polluted rivers and the extremely high levels of air pollution were part of the everyday post-Mao experience. However, by the dawn of the 21st century, China's economic growth and development patterns were spatially uneven with concentrations of foreign investment in the eastern coastal cities and Pearl River Delta in the south.

Book scope and structure

This book introduces the development, design and meaning of four purpose-built public parks through the synthesis of hybrid modernity, a new design language, and set against the backdrop of China's late 20th century process of hybrid modernization. It takes into account literature disseminated after 2010 when the author submitted the original PhD manuscript – essentially expanding from these works. The post-2010 literature magnified the knowledge deficiency on China's modern park design, especially the post-Mao characterization for the landscape architecture audience. The author's intention is to update the original manuscript and make it more accessible for a broader audience – students, practitioners and scholars interested in purpose-built public parks, design history and landscape architecture theory, particularly cultural development during post-Mao China's late 20th century urban experiment. The book re-orients this audience to China's world stature as one of the foremost long-standing civilizations. Furthermore, with the hyper-speed of change in China, this book documents four post-Mao purpose-built public parks – bearing in mind the potential for their "erasure" due to political reasons or the climate crisis.

Throughout the book, Chinese terminology has been translated into English using the *Hanyu pinyin* system, the Romanized Chinese phonetic system for *Putonghua*, China's official national language, and the currently accepted form of translation. Most of the bilingual translations (English followed by *Putonghua*) were obtained from the English literature, cross-checked through lengthy discussions with Chinese language fluent colleagues, as well as referencing the Oxford English-Chinese/Chinese-English dictionary. The book respects China's normative practice for personal names and uses the traditional sequence: the family name (usually ancient clan reference) is first and the given name second.

For the landscape architecture audience, this book traverses beyond the traditional history of park design and navigates into the transdisciplinary realm of cultural production in post-Mao China. It foregrounds the history of parks against a broad view of China's modernity and introduces the reader to modernization theory, development studies and cultural theory. The conceptual synthesis of hybrid modernization and related hybrid modernity act as the crucible for understanding the convergence of late 20th century international and local design influences, post-Mao cultural trends and production, nation-building and Chinese identity. Given public parks are generally linked to modern city-making, this book introduces the purpose-built park as a commodity for post-Mao city-making during contemporary China's vast urban experiment. From this perspective, the book expands history and theory-building in landscape architecture by introducing hybrid modernity as both a theoretical and analytical framework for situating late 20th century parks in post-socialist China. As the first English language account on the history of public parks in modern China, it is foundational. The author also introduces the "Chinese chapter" to the "international world book" on modern landscape architecture. In this light, the author endeavors to create a bridge between the literature in China and the English-speaking world, and expands theory-building in the discipline of landscape architecture and studies of the built environment.

For the non-design and non-landscape architecture audience, this book introduces the public park, one of the archetypal forms in landscape architecture praxis and research, and its historical development in both the west and modern China. It reveals innovations in park design during China's vast late 20th century urban experiment. It introduces the Chinese Picturesque as depicted in China's traditional gardens and culture, and its influence on local design, identity and cultural production. The intention for this audience is both discovery and revelatory – that this study of parks, its historical development in modern China and spatial expression in the hybrid modern late 20th century period, will provoke interest in a world civilization and post-socialist nation, once globally isolated and subsequently opened to the postmodern western world. As the first book of its kind, it can be considered as a baseline or a series of footnotes on a variety of topics that factor into a transdisciplinary and interpretative narrative for understanding the public park's evolution during China's journey to build a post-imperial nation. Alternatively, for both the non-design and landscape architecture audiences, the author hopes to provoke further transdisciplinary thinking and future studies along various or convergent lines of inquiry.

The book chapters are organized around independent topics and themes – complementing each other as a form of tapestry or textile art. Interwoven together, these discrete chapters elucidate late 20th century urban China as part of a social evolutionary process of hybrid modernization. In tapestry-making made simple, various colored threads create patterns that form the pattern and details for the pictorial image with other threads critical for the integrity of the tapestry's structure. Each of the chapters represents the various types, colors and patches of threads critical for making the tapestry's final pictorial image legible. When read as a whole, the book portrays the fabric of a multi-layered tapestry in which the chapters interweave into a complementary pattern. Individual chapters could also be read on their own, though every attempt is made to cross-reference topics. In essence, the book is not meant to portray a traditional historical synthesis and singular history of public parks. It forms a transdisciplinary narrative for understanding purpose-built public parks as part of cultural production for post-imperial and modern, and post-Mao and post-socialist China nation-building.

Chapter 2 introduces the reader to a variety of concepts, definitions and themes and characterizes modern China as a world civilization. Themes include nation-building and China's 20th century revolutionary praxis, *Chineseness* and the idea of local identity, as well as tensions between modernity and tradition. It also introduces a broad and preliminary narrative on the public park's evolution against the backdrop of China's "modern" journey beginning in the late Qing imperial dynasty circa 1840s and traversing through the Mao period. China's modern history of the public park considers political and institutional shifts, as well as introducing parallel and overlapping themes. Interwoven in this chapter is some discussion of western 19th and 20th century public parks as devices of modernization, city-making practices and urban design strategies. While the book is not intended to compare the design history of parks in China to the west, it re-visits the historical and theoretical development of public parks in the western world.

The book illuminates China as a long-lived world civilization through its garden tradition and culture with the introduction of their private garden archetype. As noted earlier, the private garden is one of two archetypes in the discipline of landscape architecture with the public park as the companion archetype. Chapter 3 foregrounds China's distinctive garden tradition and culture against a review of their pre-modern concepts on nature, folklore and mythology and religious influences. It touches on the typology of private gardens and outdoor public space in pre-modern China, major influences and inspirations for the traditional private garden – Scholar Gardens, or Gardens of the Literati, *wenren yuanlin* – amateur gardens designed by and for retired officials from China's imperial court. It highlights garden-making as an extension of China's classical arts, or "three-dimensional painting", that evolved over several centuries and includes a discussion of West Lake's iconic landscape and its significance to cultural development in both the imperial literati period and early 20th century modern city-making practices. The author deems this overall imagery as the traditional "Chinese Picturesque" genre and Chapter 3 summarizes the design language, grammar and vocabulary deployed in the garden's spatial forms and acknowledges a formidable body of scholarly work in both the Chinese and English language exists on Chinese traditional gardens. China's dynastic time, Figure 1.4, represents the normative chronology accepted by historians. The Chinese Picturesque garden tradition and culture is a critical component to the meaning of "local" identity and local design influence in the theoretical construction of hybrid modernity.

Chapter 4 reviews various themes and trends expressed in the arts – getting at the essence of what was happening culturally "inside" during the post-Mao period. The *zeitgeist* and artists' attitudes shifted as rapidly as China's late 20th century accelerated urbanization and were responsive to the central government's changing meaning of an

Ancient China	Xia		2100 - 1800 BCE
	Shang		1700 - 1027 BCE
	Western Zhou		1027 - 771 BCE
	Eastern Zhou 700-221 BEC	Spring & Autumn Period	770 - 476 BCE
		Warring States Period	475 - 221 BCE
Early Imperial China	Qin		211 - 207 BCE
	Han		206 BCE - 220 CE
	Three Kingdoms		220 - 280
	Jin		265 - 420
	Sixteen Kingdoms		304 - 439
	Southern and Northern		420 - 589
Classical Imperial China	Sui		581 - 618
	Tang		618 - 907
	5 Dynasties and 10 Kingdoms		907 - 960
	Song 960 - 1279	Northern Song	960 - 1127
		Southern Song	1127 - 1279
Later Imperial China	Yuan		1271 - 1368
	Ming		1368 - 1644
	Qing		1644 - 1911
Modern China	Republic of China		1912 - 1949
	People's Republic of China [New Era 1982]		1949 -

Figure 1.4 General chronology of China based on Fairbank (1957), Clunas (1997) and Padua (2014), graphic by author

1 Living Water Park   Chengdu, Sichuan
2 Zhongshan Shipyard Park   Zhongshan, Guangdong
3 Jinji Lake Master Plan   Suzhou, Jiangsu
4 Southern Scenic Area, West Lake   Hangzhou, Zhejiang

Figure 1.5 Context map with locations of the four case study parks, graphic by X. Liu

"open" society. The chapter also touches on cultural development in the Mao era, the emergence of the profession of landscape architecture and institutional changes in education in post-Mao China. The relationship of landscape architecture education to the development of the profession of landscape architecture in China is no doubt a major study in itself. However, an institutional glimpse of design education was too important to neglect entirely. Chapter 4 touches on cultural influences, international landscape architecture praxis, design trends and themes happening "outside" China when the nation opened to the postmodern world.

Chapter 5 reveals hybrid modernity through the case study exploration of four pur-pose-built projects (Figure 1.5), namely the Living Water Park, Zhongshan Shipyard Park, and the projects from master plans of Jinji Lake and West Lake's Southern Scenic Area, in the four respective Chinese secondary cities: Chengdu, Sichuan Province; Zhongshan, Guangdong Province; Suzhou, Jiangsu Province; and Hangzhou, Zhejiang Province. The case study analysis of these four parks serves as the empirical evidence and schema for the synthesis of hybrid modernity – a theory for understanding post-Mao China. Themes, design language, influences and concepts from previous chapters are deployed in the analysis and situate these four parks within the larger context of the discipline of landscape architecture. Chapter 5 also reflects on the evolution of landscape architecture praxis during China's late 20th century.

The book's sixth and final chapter speculates on the vitality of China's 21st cen-tury landscape architecture praxis, particularly in light of the massive environmental degradation and challenges generated by its late 20th century urban experiment and the agency of a single-party nation. It discusses new designed landscapes (greening, parks and landscape systems) and government initiatives in the four secondary cities since the four parks were completed. It also touches on China's 21st century initiatives related to rural-urban planning, conservation and design of their built and natural environments, especially the Sponge City national pilot initiative. It highlights China's top-down governance and suggests their "green" revolution circa 2007 triggered the shift from "hybrid" to "ecological modernization" when the government re-oriented away from Deng's "growth first" unregulated development to former President Hu Jintao's "people first" strategy. The latter is based on environmental restoration, conservation of natural resources, energy efficiency – fundamental to China's "ecological civilization construction" and the nation's health and well-being. It touches on current President Xi Jinping's "Beautiful China", a vision grounded in the reduction of the nation's environmental pollution, and restoration and conservation of China's natural resources. Given the "wicked" challenges of the contemporary Anthropocene era, the final chapter postulates China as a major 21st century world leader in environmental stewardship by actively dealing with the climate crisis through resilience and sustainable practices. In this 21st century period of ecological modernization, China self-identifies as a living laboratory where allied disciplines working in the built and natural environments will continue to flourish, especially landscape architecture.

References

Anderson, B. (1991) *Imagined Communities: Reflections on the Origin and Spread of Nationalism*. London, UK, Verso.

Appadurai, A. (1990) Disjuncture and Difference in the Global Cultural Economy, *Theory, Culture, and Society*, 7(2), 295–310.

Baljon, L. (1999) *Designing Parks: Examination of Contemporary Approaches to Design in Landscape Architecture*. Amsterdam, Netherlands, Architectura and Natura Press.

Bhabba, H. K. (1994) *The Location of Culture: The Commitment to Theory*. London, UK, Routledge.

Bull, C. (2006) *New Conversations with an Old Landscape: Landscape Architecture in Contemporary Australia*. Mulgrave, Victoria, Australia, Images Publishing.

Byrd, W. & Lin, Q. eds. (1990) *China's Rural Industry: Structure, Development, and Reform*. Oxford, UK, Oxford University Press.

Campanella, T. (2008) *The Concrete Dragon*. New York, NY, Princeton Architectural Press.

Clunas, C. (1997) *Art in China*. Oxford, UK and New York, NY, Oxford University Press.

Connolly, P. (1995) '101 Ideas' about Big Parks, *Kerb: Journal of Landscape Architecture*, 1., 20–35

Crane, B., Albrecht, C., McKay Duffin, K.. & Albrecht, C. (2018) China's Special Economic Zones: An Analysis of Policy to Reduce Regional Disparities, *Regional Studies, Regional Science*, 5(1), 98–107.

Cranz, G. (1979) The Useful & the Beautiful: Urban Parks in China Landscape, *Landscape Journal*, 23, 3–10.

Cranz, G. (1991) Four Models of Municipal Park Design in the United States. In: *Denatured Visions: Landscape and Culture in the Twentieth Century*, Wrede, S. & Adams, W. (eds.). New York, NY, Museum of Modern Art.

Czerniak, J. & Hargreaves, G. eds. (2007) *Large Parks*. Princeton, NJ, Princeton Architectural Press.

Davies, H. & Raskovic, M. (2017) *Understanding a Changing China*. London, UK and New York, NY, Routledge.

Davis, D. S. (1995) Introduction: Urban China. In: *Urban Spaces in Contemporary China: The Potential for Autonomy and Community in Post-Mao China*, Davis, D. S., Kraus, R., Naughton, B., & Perry, E. (eds.). Washington, DC, Woodrow Wilson Center Press.

Dirlik, A. (2002) Modernity as History: Post-Revolutionary China, Globalization and the Question of Modernity, *Social History*, 27(1), 16–39.

Dong, Y. D. (1999) Defining Beiping: Urban Reconstruction and National Identity, 1928–936. In: *Remaking the Chinese City: Modernity and National Identity, 1900–1950*, Esherick, J. W. (ed.), Honolulu, HI, University of Hawaii Press.

Esherick, J. W. (1999) Modernity and Nation in the Chinese City. In: *Remaking the Chinese City: Modernity and National Identity,1900–1950*, Esherick, J. W. (ed.), Honolulu, HI, University of Hawaii Press.

Fairbank, J. K. ed. (1957) *Chinese Thoughts and Institutions*. Chicago, IL and London, UK, The University of Chicago Press.

Fan, C. (1995) Of Belts and Ladders: State Policy and Uneven Regional Development in Post-Mao China, *Annals of the Association of American Geographers*, 85(3), 421–449.

Friedmann, J. (2005) *China's Urban Transition*. Minneapolis, MN, University of Minnesota Press.

Gao, M. (1998) *Inside Out: New Chinese Art*. Berkeley, CA, University of California Press.

Gao, M. (2011) *Total Modernity and the Avant-Garde in Twentieth-Century Chinese Art*. Cambridge, MA, MIT Press.

Hayden, D. (1995) *The Power of Place: Urban Landscapes as Public History*. Cambridge, MA, MIT Press.

Jackson, J. B. (1970) *Selected Writings by J. B. Jackson*. Amherst, MA, University of Massachusetts Press.

Jencks, C. (1978) *The Language of Post-Modern Architecture*. New York, NY, Rizzoli Press.

Jencks, C. (2002) *The New Paradigm in Architecture: The Language of Post-Modernism*. New Haven, CT, Yale University Press.

Koolhaas, R. (2002) *Great Leap Forward*. Cambridge, MA, Taschen.

Liu, J. (1971) Mao's 'On Contradiction', *Studies in Soviet Thought*, 11(2), 71–89.

Lu, D. (2006) *Remaking Chinese Urban Form: Modernity, Scarcity and Space, 1949–2005*. London, UK and New York, NY, Routledge.

Ma, L. J. C. & Wu, F. (2005) *Restructuring the Chinese City*. London, UK and New York, NY, Routledge.

Meyer, E. (1991) The Public Park as Avante-Garde, *Landscape Journal*, 10(1), 16–26.

Ouroussoff, N. (2008) Lost in the New Beijing: The Old Neighborhood, 23 July 2008, *New York Times*.

Padua, M. (2003) Industrial Strength, *Landscape Architecture*, 93(6), 76–85 & 105–107.

Padua, M. (2004a) Teaching the River, *Landscape Architecture*, 94(3), 100–109.

Padua, M. (2004b) Future Scale, *Landscape Architecture*, 94(8), 106–115.

Padua, M. (2006) Touching the Good Earth, *Landscape Architecture*, 96(12), 100–109.

Padua, M. (2007) Designing an Identity: The Synthesis of a Post-traditional Landscape Vocabulary in Hong Kong, *Landscape Research*, 32(2), 255–272.

Padua, M. (2008) A Fine Red Line, *Landscape Architecture*, 98(1), 92–99.

Padua, M. (2012) This Way Shanghai, *Landscape Architecture*, 102(12), 54–65.

Padua, M. (2013a) Bridge to Somewhere Else, *Landscape Architecture*, 103(2), 80–87.

Padua, M. (2013b) Triptych by the Sea, *Landscape Architecture*, 103(2), 98–107.

Padua, M. (2014) New Cultures and Changing Urban Cultures. In: *New Cultural Landscapes*, Roe, M. & Taylor, K. (eds.). London, UK and New York, NY, Routledge.

Rinaldi, B. (2011) *The Chinese Garden: Garden Types for Contemporary Landscape Architecture*. Basel, Switzerland, Birkhauser.

Schenker, H. (1995) Parks and Politics During the Second Empire in Paris, *Landscape Journal*, 14(2), 201–219.

Skinner, G. W. (ed.) (1977) *The City in Late Imperial China*. Stanford, CA, Stanford University Press.

Tate, A. (2001) *Great City Parks*. New York, NY, E & FN Spon Press.

Thompson, C. W. (1998) Historic American Parks and Contemporary Needs, *Landscape Journal*, 17(1), 1–25.

Waldheim, C. ed. (2006) *Landscape Urbanism Reader*. New York, NY, Princeton Architectural Press.

Wang, H. (2003) *China's New Order : Society, Politics, and Economy in Transition*. Cambridge, MA, Harvard University Press.

White, S. K. (1989) *The Recent Work of Jürgen Habermas: Reason, Justice and Modernity*. Cambridge, UK, Cambridge University Press.

Wu, P. & Gaubatz, P. (2013) *The Chinese City*. London, UK and New York, NY, Routledge.

Wu, Y., Luo, J., Zhang, X. & Skitmore, Martin (2016) Urban Growth Dilemmas and Solutions in China: Looking Forward to 2030, *Habitat International*, 56, 42–51.

Yu, K. & Padua, M., eds. (2006) *The Art of Survival: Recovering Landscape Architecture*. Mulgrave, Victoria, Australia, Images Publishing Group Pty Ltd.

Yu, K. & Padua, M. (2007) China's Cosmetic Cities: Urban Fever and Superficiality, *Landscape Research*, 32(2), 225–249.

Zhang, X. (1997) *Chinese Modernism in the Era of Reforms*. Durham, NC, Duke University Press.

2 Navigating modern(s) and the park in modern China

The inter-related concepts of modernization, modernity and modernism, or "moderns", help inform the synthesis of hybrid modernization and related hybrid modernity – a theoretical construct for China's 1978–2001 post-Mao period. The convergence of these concepts serves as the broad canvas for navigating modern China and the public park. It orients the reader to modern China, an extremely complex topic on its own, and discusses the meaning of parks in China's Mao and pre-Mao periods. This in turn informs the book's focus on the post-Mao late 20th century public park. As a world civilization continuously operating for thousands of years, China's political and spatial dimensions are also considered in the discussion. The author recognizes the "thickness" of the literature on modernization, modernity and modernism, especially as it relates to China, and the intention is not to be reductive or simplistic. However, it offers a transdisciplinary perspective for understanding the public park's emergence in modern China.

The chapter is organized into two constituent parts where one informs the other. The first section explores modernization, modernity and modernism, broadly and within the context of China; and it sets the context for the second part of the chapter – navigating the public park, its emergence and evolution in modern China. It touches on early modernity and the introduction of European Renaissance gardens by the Roman Catholic Jesuit priests at the imperial gardens located in the Old Summer Palace near Beijing. It discusses more deeply the emergence of parks for foreign consumption in China's late Qing dynasty and its meaning in cities during the short-lived modern Republic of China, particularly as agents for nation-building. The chapter concludes with a discussion of the socialist production of "greening, *luhua*" and people's parks, *renmin gongyuan*, within Mao's era of "extremes" and "contradictions".

Modernization is both process and action-oriented, and often involves the updating of technology and other material concerns. As noted in the introductory chapter, modernization theory usually falls within the purview of scholars in the social sciences and humanities. This work is framed as a broader social process driven by economic policies and nation-building practices. A basic dictionary definition of modernity states the phenomena as having to do with the quality or state of being modern. Some cultural and social theorists have positioned it through the binary critique of modernity versus tradition – or modernity as a break from the past. Harrison and Wood (1992) add to this definition of modernity, the notion of a shared social experience with the awareness of change and adaptation to change. Modernism refers to a mode of expression, or type of style or workmanship characteristic of modern times. Within the context of cultural production, modernism in the western world has been argued

from several vantage points, wherein scholars have taken into account certain variables. The tripartite synopsis below touches on themes pertinent to understanding
modern China. Taken together, these concepts, themes and introduction to modern
China paint a backdrop for understanding the public park in the period that spans
the 1850s through the late 1970s, respectively the late Qing dynasty through the Mao
era. As Wang (2003, p. 160) notes, "Chinese New Enlightenment thinking is without
a doubt the most influential of all ideologies of modernization".

Modernization: social construction and nation-building

Studies of modernization have charted the shape of a nation-state's social structure
and culture, framed within an economic context. Modernization theory has played a
major role in fields such as development studies, cultural studies, sociology and planning, but it has received little attention in landscape architecture. This is partially due
to the fact that works produced by landscape architects in the mid to late 20th century
were often treated as a form of "high art" (Treib 1993; Johnston 2006). Major exceptions to this approach include works by Ian McHarg (1969), J. B. Jackson (1970), as
well as Lawrence Halprin (1969) and Clare Cooper Marcus (1974). McHarg (1969)
pioneered landscape studies at the regional scale using multiple layers of mapped data
and the suitability method. Jackson and his followers' work have dealt primarily with
"undesigned" vernacular or everyday landscapes not necessarily created by professional designers but created as a result of local culture or agrarian practices. Halprin
and his wife, Anna Halprin, along with Jim Burns (Halprin & Burns 1974) pioneered
participatory workshops where community members actively engaged in the design
processes. Clare Cooper Marcus pioneered the importance of social and behavioral
factors in designed outdoor environments with post-occupancy evaluation, and later
in her career, therapeutic gardens and the salutogenic approach. Simultaneously, concepts of "sense of place" and "place-making" emerged as vital social and community
variables for defining "quality of life" in landscape architecture and urban planning as
a result of work by William Whyte (1980), Jan Gehl (1971) and Jane Jacobs (1961).

By the late 1980s and 1990s, the Urban Land Institute raised the importance of
open space systems, public parks, designed landscapes and art in public places as
agents for economically revitalizing cities. Additionally, the concept of landscape
urbanism emerged in the 1990s, whereby open spaces, designed landscapes or the
horizontal spaces between buildings open to the sky became the primary form-driving
factor for remaking cities in the western world (Connolly 1995; Waldheim 2006). It
spawned the importance of the tectonics of the outdoor designed environment as having a dynamic and performative function, particularly in the design and planning of
urban public infrastructure: greenways, vegetation as water-absorbers, cleansers and
living machines, and adaptations of conventional subterranean stormwater management systems to naturalized on grade or above ground systems. Except for landscape
urbanism, these previous works largely contributed to setting the foundation for cultural, social and community issues in landscape architecture in the western world. This
book attempts to expand the discipline and scholarly research in landscape architecture in terms of theory-building and 20th century design history, especially the public
park in post-Mao China.

Contemporary discussion of modernization has its roots in work appearing in
the period after World War II (WWII). The 1950s and 1960s were marked by the

emergence of new nations in Africa and Asia and a massive push toward economic development. The idea of modernization was used then to describe the social side of economic development and nation-building. It asserted that "social modernization" was a necessary condition for economic development, and that the process of industrialization created further social modernization (Levy 1966). This led to the idea that a "logic of industrialization" existed and inexorably drove industrial societies toward similar social structures, norms and values (Kerr et al. 1964). Karl Marx's interpretation of modernization in Europe did not attribute it to the industrial revolution and technological change. Instead, Marx's interpretation based technological change and innovation on a prior change in social institutions including mores and social behavior (Avineri 1968). Later, a group of Harvard sociologists led by Alex Inkeles celebrated the idea that the industrial factory is a "school for modernization" (Inkeles & Smith 1974). Later, the link between social change and economic development were interconnected, especially as western society evolved (Levy 1966; Inkeles & Smith 1974). The arguments went so far as to assert that economic development was impossible without social changes.

The resemblance to American values was not lost on scholars (Said 1977) and a substantial body of work soon refuted the logic of industrialization arguments and the modernization theories of the 1960s. The critique largely came from the political left, and various forms of socialist development often were cited as counter-examples to the modernization arguments – originally framed from the standpoint of a market economy or capitalist society. Social theorists such as Fernando Henrique Cardoso (later president of Brazil) noted the underdevelopment of the third world was a product of global history and created by the same colonial and post-colonial policies that helped to make the industrial world rich (Cardoso 1979). A world order evolved with a tier of dominant industrial countries at the center and the poorer countries of the third world occupying the periphery (Frank 1969). In this view, economic change did not automatically move societies toward universal modernization. Economic development was essentially independent of modernization, and it could proceed according to models that had little to do with the experiences of 19th and 20th century Western Europe.

This debate went on in a variety of forms in several different disciplines during the 1960s and 1970s with economists and economic historians discussing the role of colonialism and world trade in the economic development of Europe (Braudel 1974). Modernization theory was contested among social scientists, and opposition to it spawned the ideas of "dependency theory" and later "world systems theory" (Cardoso 1979; Wallerstein 1974). The debate over modernization shifted by the 1980s when social theorists such as Anthony Giddens (1990) recast the idea of modernization as a broader concept taking on different forms in different societies. This argument found a ready audience among researchers, socio-cultural scientists and policy-makers looking at development in an international context. The idea of international influences on the modernization of societies reappeared in a form that had been stripped of the determinism expressed in the early views of modernization.

Appadurai (1990), Bhabba (1994) and others argued social and economic change could simultaneously be influenced by global forces and remain culturally, spatially and geographically idiosyncratic. They viewed the process of change as embedded within culture, but the societies in which change is taking place are not isolated from the larger world. In the opposing position, globalization can be a major force for change without compromising the uniqueness of local cultures. The results of change are not determined solely by international forces, they also reflect local culture. The process of modernization and the form that modernity takes thus may be specific to individual societies.

Appadurai (1990) described this idea as the existence of "alternative modernities". Ideologically, the emphasis is the fact that China and India are world civilizations with their own cultures and spatial structures. Hence, these civilizations need not be examined through the lens of western perspectives on modernization. Essentially, the western binary discourse of comparing China as a "third world" nation to the "first" or westernized world and similar dichotomies, i.e. developing/developed, industrial/post-industrial, should no longer be the focus. The concept of hybrid modernization and related hybrid modernity in post-Mao China diverges away from this dichotomous thinking.

Modernity, globalization and identity

The view of modernization and modernity that emerged from the 1980s and 1990s literature opened up a variety of new questions. Marxist sociologists such as Manuel Castells (1989) factored in the important role of the media in promoting global modernity with the term "space of flows" – the emerging global social order as a result of technology, communications and mass media. Discussion of modernization, globalization and post-colonial identity evolved into a major concern in the field of cultural studies (Appadurai 1990; Bhabba 1994; Dirlik 2002). The concept of hybrid modernization draws from this work.

The notion of globalization was the *zeitgeist* of the 1990s through the turn of the 21st century. For the last thirty years, it has grown to be accepted, debated and highly contested among social scientists, cultural theorists, economic historians, community activists, and individuals in societies around the world. As we now know it, globalization emerged as a result of technological advances in communications, including the internet and worldwide web, global competition and multinational corporations operating on a world scale, outside the boundaries of nation-states. For some, globalization appeared to replace modernization and concerns were raised that globalization in its extreme form could influence the demise of national economies and potentially nation-states. However, contemporary researchers interested in modernization place more emphasis on the independent power of ideology and the production of culture by local societies. They see culture and human identity as phenomena that are created through the interplay of global forces and local history and culture.

Sliding into the turn of the 21st century, social scientists and cultural theorists noted the polarities of modernity and tradition as inadequate to describe the phenomenon of modernization or globalization as it was unfolding in the contemporary world (Bhabba 1994; Chakrabarty 2000). Modernization does not necessarily spell the death of tradition. However, it does call into question the concept of tradition. In this light, the idea of tradition itself is seen as a cultural product developed in interaction with the forces of modernization. Cultural theorists question whether the concept of tradition is a "Eurocentric" idea that imposes an identity on third world people primarily as "the other" (Bhabba 1994; Appadurai 1990). In this book, the research on late 20th century China builds from the notion that identity is produced through the interplay of global and local influences, and nation-building. "Tradition" has been redefined from a simple historical artifact to a phenomenon that is a contemporary creation formed out of both local and international influences. The concept of tradition does not exist without the idea of modernity, and modernity itself emerges as a deeply ambiguous concept in this context.

Identity is a creation built on foundations of locality and globalization, just as modernity and tradition are polarities that have both been created by larger forces. The work of Benedict Anderson (1983) has been particularly important in contemporary work on location and identity. Anderson argues that identity itself is largely mythological, a cultural artifact produced for various social and political purposes. He points out that actual community is restricted to a handful of people who are able to interact directly. Despite this fact, individuals typically see themselves as belonging to larger regional and national communities whose members they have never met, never will meet, and who may not even exist. In post-colonial cases, this might take the form of an abiding attachment to a "mother country" that the individual has never seen and knows only through oral history and social mythmaking or to an ancestral past believed to have existed before historical record. For example, South Asian Indian civil servants who have never traveled to the United Kingdom may still think of England as a place where their values are rooted, and Chinese in Hong Kong may trace their village lineage back a dozen generations to a place and time on mainland China that have left no historical trace. As Anderson (1983, p. 6) notes, "Communities are to be distinguished, not by their falsity/genuine-ness, but by the style in which they are imagined".

Anderson (1983) argues that virtually all forms of community have this imagined, socially constructed character. This is particularly true of local and national identities, and nation-building in the case of China, especially during the 19th and 20th centuries. Anderson (1983) builds on work in intellectual history that treats nationality as a constructed political identity and expands further, noting concepts such as ethnic community are highly politicized forms of identity. Social identities are created for political ends, as a means of mobilizing individuals under the banner of nationalism. Seen through this lens, the concept of modernization refers to a process of reconstructing identities in response to the influence of global forces. Local identity, national identity, and supra-national identities all are created by social and political forces acting at different levels. This has been a very compelling formulation of the concept of modernization, but it is peculiarly indifferent to the physical reality of place.

Hybrid modernization: expanding modernization theory and alternative modernities

The author has deemed "hybrid modernization", a variation of modernization theory, more useful in understanding China's late 20th century and post-socialist society. It is built on the foundation of contemporary work on modernization. It embraces Anderson's notion that national identity is essentially a mythological construction largely serving political purposes. The idea of modernity also is a social and cultural product, and it is strongly influenced by ideologies of progress that have their roots in the French Enlightenment. Globalization has played a major role in implanting these ideas within China and much of the rest of the world. However, locality also has a concrete component rooted in the realities of place and the history of a place. It's vital to note China in 1978 was essentially impoverished, recovering from the Cultural Revolution and finding its way out of global isolation while it awakened to a postmodern world.

The interpretation of the history of a place may be deeply influenced by the global forces of modernization, but the place nonetheless exists and generally has some historical record. Recent history, in particular, often has been part of the direct experience of the people in that place. And it forms one element of identity, an element that is far less mutable than the forms of identity discussed by Anderson (1983).

Landscape architects and planners often discuss "place-making" as a means of creating local identity. In practice, place-making is a process of creating a local identity but the process is constrained by the physical form and history of a place. Places have a physical form and a set of people who have experienced their physical reality, and the definition of a place is constrained by these facts. The identity given to a place is not solely determined by its physical form, but it is not completely independent of physical form. The vast indoor waterpark with swimming areas and a wave-making machine for surfing in the Mall of America shopping center in Minnesota state attracts many of the uses of an outdoor water recreational space; teenage boys surf on artificially generated waves and teenage girls display sun-studio tans in mid-winter. Nonetheless, the shopping center remains a shopping center, the pools, water slide and wave-making machine are constructed features, and many of the local shoppers can remember when the whole thing was "the Met", the former sports venue and baseball field, Metropolitan Stadium in Bloomington, Minnesota, just outside Minneapolis.

The socio-economic dimension and benefits of public parks and green open space systems have been folded into the notion of place-making in the public realm. Since the 1980s or so, American policy-makers have acknowledged the important relationship between the image of the urban fabric of their cities and towns, and economic revitalization. Local city and town officials in the USA, UK and EU understand the value of trees, vegetation and pedestrian friendly environments and public open space systems for community-building and as symbols of open and democratic societies. Community stakeholders and municipal administrators in American cities work with landscape architects and planners to engage their local communities to improve their community's sense of place with new public open space, especially as tools for economic revitalization.

The concept of hybrid modernization retains the idea of place as an element of both locality and international influences in post-Mao China. Local influences on a hybrid modern form include the elements of identity and the physical, economic and historical influences of place. Local forces thus are a result of a complex interaction among the physical realities of place and its experienced history, the culturally and socially influenced interpretation of that history, and the social definition of the locality as different from some global modernity.

Giddens (1990), a social theorist, investigated society as emblematic of a particular nation's progress, while cultural theorists Bhabba (1994) and Appadurai (1990) examined the cultural development of a particular society as the primary influence on national identity. Bhabba (1994) and Appadurai (1990) contend particular societies and their cultures evolve in response to international influences or so-called globalization. Appadurai's notion of alternative modernities also takes into account the geo-spatial and cultural development of China and India as major world civilizations that should be evaluated on their own. The concept of hybrid modernization and related hybrid modernity aligns with this trajectory and deepens the argument. Anderson (1983) interprets national identity and its re-imagining as highly influenced by the impact of media with the result of myth-making.

Hybrid modernization emerges out of this socio-cultural discourse for interpreting China's identity as a nation and world civilization with a unique development trajectory. It expands on modernization theory and is shaped by local forces in combination with the international influences of globalization, nation-building and national identity. Identity has been created in the fashion that Anderson (1983) describes – a process of mythmaking, often politically charged. The idea of hybrid modernity is particularly useful in the Chinese setting. Nation-building, local cultural identity and global influences on society have been deeply intertwined in post-Mao China, especially after global isolation. Hybrid modernization highlights the way these forces interact in shaping late 20th century China's society in their New Era. At the same time, China has a deeply engrained and deeply mythologized national identity that is fundamentally different from local identity. At points during the 20th century, the two were directly at odds with each other. The mingling of locality and modernity combine with this larger "Chinese" identity to shape key elements of the design of the new purpose-built parks in late 20th century China. These forces are not entirely independent of each other and they have their own histories. More distinctively, they are manifested in the hybrid modern design genre represented in the book's four case study parks.

Re-visiting modernism (international) genre

Modernism in the arts from the western perspective, and among historians, has largely been viewed as a style or movement that emerged in the late 19th century when artists and designers were purposefully seeking a break from the past. Their objective was to innovate and experiment with materials, forms and techniques that reflected the culture of the times or so-called modern society. It is widely accepted that modernism spans the period between the late 19th century through the 1970s or so. The synopsis below offers chronological highlights of various genres that fall under the larger rubric of modernism. The overall objective is to re-orient the reader to modernism in the arts, historically, thematically and geographically, as well as to help situate modern China and international cultural influences that may have made their way there.

The period of modernism in the arts has been defined historically by the period beginning in the late 19th century through the 1970s. Several genres and movements first emerged in particular locations in western Europe, Great Britain, Russia and the Americas during this period. The early decades of the 20th century were tumultuous with changes caused by industrialization and modernization processes in the western world. Works by artists, poets and intellectuals were responding to this "machine age" or period of modernity, as well as to the socio-cultural milieu created by the turmoil of the "Great War" or World War I (WWI). Some movements and genres were influential, crossing political boundaries throughout Europe and making their way across the Atlantic Ocean, with a few that were geo-spatially specific. "Constructivism", and later their government-sanctioned "Socialist Realism" in Soviet Russia, are some examples of geo-spatial specific modern genres. The latter made its way to China in the Mao era.

In the visual arts, especially the medium of painting, the notable modern art critic Greenburg (1989) viewed modernism as self-critical and saw its roots in Kantian philosophy. For Greenburg, modern painting or modern art was less a break with tradition but more a devolution from an earlier period perceived in a positive manner. Greenburg notes (1989, p. 9): "Modernist art develops out of the past without gap or break, and wherever it ends up it will never stop being intelligible in terms of the continuity of art".

Widely accepted as the precursors to modernism in the disciplines of fine arts, applied and decorative arts, are the British Arts and Crafts movement, circa 1880–1920, and western European Art Nouveau movement, circa 1890–1910. Leaders of the Arts and Crafts movement were responding to England's Victorian era and industrialization, poor design and quality of manufactured goods for domestic use, changes in society and the rise of the consumer class. Their view was there should be unity in the arts and they rejected the idea that there were distinctions between fine arts and the applied arts. They advocated social awareness, egalitarianism and their focus was on creating an appreciation of good design, individual craftsmanship and traditional techniques for all members of society. In this context, for example, both the makers and users of furniture would gain a sense of pleasure (Cumming 1997).

The Art Nouveau movement is seen as an outgrowth of the Arts and Crafts movement. Graphic artists, architects, furniture designers, jewelry designers, and others were swept up in the Art Nouveau movement. Examples of Art Nouveau, and their curvilinear organic forms, were inspired by plant and floral forms from nature. The iron and glass pavilions at a few remaining Metro stations in Paris, France designed by French architect, Hector Guimard, are living examples. Art Nouveau was a broad European movement also known as *Jugendstil* in Germany, *Sezessionstil* in Austria, *Stile Liberty* in Italy and in Spanish Catalonia, *modernista* (Bonnet 1992). Both the Arts and Crafts and Art Nouveau movements found their way to the USA and elsewhere, and were viewed as largely utopian.

Before the emergence of the Arts and Crafts and Art Nouveau movements, Édouard Manet, an early Impressionist painter, circa 1860s in Paris, France, was known to break from the style of the Old Masters. His looser style suggests early modernism. Eventually, as modern society evolved and Europe industrialized, cities throughout Germany and Austria in the first decade or so of the 20th century, the Expressionism movement emerged. Wassily Kandinsky, a Russian artist based in Munich, Germany, and Franz Marc, a Munich native, were considered leaders of Expressionism. It was a movement concerned about the fate of humanity and modern society. Works of this genre were considered more inward and spiritual, critical of society and highly abstract (Harrison & Wood 1992).

Between 1907 and 1914, Cubism, a major art movement, was led by Pablo Picasso and Georges Braque in Paris, France. Viewed as innovative and radical, Braque, Picasso and others broke with the tradition utilized by the Old Masters, especially their use of perspective, modeling, foreshortening and illusion of depth, and copying nature. Instead, Cubist painters tended to emphasize the flatness or two dimensionality of the painting surface; and their works experimented with geometric forms and abstract imagery with multiple vantage points (Cooper 1971). Cubist visual artists and their abstract imagery were apolitical, and were primarily challenging the limits of traditional representational techniques rather than art itself.

The more radical "Futurism (*Futurismo* in Italian)" emerged as an avant-garde art movement in Milano, Italy, circa 1909, via the "Futurist Manifesto (*Manifesto del Futurismo* in Italian)" and influenced Italy's political discourse (Taylor 1968). Appropriating from Marx and Engels' 1848 Communist Manifesto, Italian poet Filippo Tommaso Marinetti, author of the Futurist manifesto, introduced this novel conceptual practice to the art world. It was the first moment in history when artists utilized the manifesto as a way of communicating their ideologies with a goal to influence individuals in society. Later, polemical written manifestos accompanied, or even preceded, actual works of art.

Considered radical, Futurists celebrated progress, radical change, advanced technology or the so-called "machine", as well as speed, youth, violence and modern urbanism. They perceived themselves as activists who rebelled against antiquated Italian art, the stagnated Italian culture of the bourgeois society. Marinetti and the early Futurists, Umberto Boccioni and Carlo Carra, concurred that artists and writers should turn their backs on the conventions of the past and embrace the vital, noisy urban life of the industrial city (Taylor 1968). Futurists were responding to Italy's social and political crisis of the time and concerned with the nation's social and cultural development, especially nation-building. Italian Futurists believed Italy could garner a new image as a world power and create a new Italian culture by erasing their past. Creative artifacts by Futurist visual artists tended toward dynamic and kinetic representations, given their interest in speed, motion and progress of the so-called machine age. Eventually, the ideology of Futurism influenced the rise of political fascism in Italy. Futurism also influenced "Dada", the first authentic conceptual art movement that emerged in Zurich circa 1916. Dada later spread to Paris, Barcelona and New York and stemmed from an anti-art and irrationalist perspective (Harrison & Wood 1992).

Like the Futurists, Dada artists sought to overturn bourgeois perspectives on the art world. The inception of the Dada movement has been attributed to artists' revolt against the traditional art and bourgeois art market, anger with the violence and destruction created by WWI and a "world gone mad". While the Dadaists were influenced by Marinetti's Futurist ideology, their movement was considered more radical with anti-art, anti-aesthetic and nihilistic perspectives. The Dada premise was one of protest – questioning art, society and politics. The terrors of WWI transformed their perspective – life was considered meaningless and somewhat dystopic. Dadaists presented this ideological perspective through performances and creative artifacts representing everyday life that mocked the pretensions of nationalist discourses and traditional art (Bonnett 1992).

As the Dada movement declined, the Surrealism movement emerged in 1920s Paris and was practiced throughout western Europe and the USA through the 1960s. The French poet, Andre Breton, wrote and published the Surrealist Manifesto (*Le Manifeste du Surréalisme*) circa 1924 (Taylor 1968). Breton portrayed Surrealism as a revolution to hundreds of years of the so-called use of the rational mind and the oppression it created. Citing Sigmund Freud's work in the manifesto, Breton declared it was vital for individuals to reunite their conscious mind with their unconscious and subconscious mind. Imagery by Surrealist painters, Salvador Dali, Joan Miro and others depicted were drawn from fantasy, dreams and myths directly relating to their self-explorations of the subconscious. Where the Dadaists worked individually and through action (protests, confrontation and spontaneous performances), the Surrealists were considered more positive, worked collectively, and as a result, were considered less innovative than the Dadaists.

In parallel with the Dada and Surrealist movements, Russian avant-garde artists and architects were engaged in the ground-breaking movement known as Constructivism soon after the Russian Revolution of 1917. Artists from various fields (visual, graphic, applied) were driven by a passion for "making", literally constructing and experimenting with materials and form (Gough 2005). Constructivists were in favor of their practice serving a social purpose – redefining a new revolutionary society in the post-czarist period under Lenin, Soviet Russia's first head of state. Between 1917 and 1921, constructivism flourished, especially in the newly established schools of art and design (Gough 2005). Once Joseph Stalin took over as the Soviet leader after Lenin's death, works by Russian avant-garde artists would be censored and replaced by the 1932 official government-sanctioned art style known as "Socialist Realism" (Reid 2001). Mao's communist regime later embraced Soviet Socialist Realism as part of China's wholesale institutional transformation to the Soviet model.

Across the Atlantic Ocean in the Americas, Diego Rivera, artist and known communist, and his contemporaries José Clemente Orozco and David Alfaro Siquieros, led the Mexican Renaissance during the period between the two world wars (Charlot 1965). Their work was part of a government-sponsored program that responded to the Mexican Revolution (1910–1920), liberating them from Spanish colonizing influences. In using the Italian fresco technique for painting murals, the resulting imagery by these artists revitalized and redefined Mexico's cultural identity. Rivera chose to have his work displayed in the public realm with his large-scale murals on buildings in public spaces depicting imagery of Mexico's pre-Columbian heritage, revolutionary imagery and national narratives of independence. This work apparently influenced the modern art genre known as Social Realism that emerged in response to the 1930s Great Depression in the USA. The program also served as a model for the Public Works of Art Project (PWAP) and later Works Progress Administration (WPA). Both were launched by President Franklin D. Roosevelt (FDR) as part of the New Deal, a federally sponsored program that was the USA's response to the Great Depression circa 1929–1939, and provided job opportunities to unemployed Americans including artists and writers. In the USA, Social Realism has also been categorized within the context of "regionalism", a type of 1930s American art that utilized a narrative style depicting the humanistic conditions of America's Great Depression (Harrison & Wood 1992). Following WWII, New York City became the world capital for the new modern painting genre known as Abstract Expressionism. This represented a major geographic and cultural shift to the USA – steering away from Paris, France and western Europe.

The radical shift to New York in the art world was highly significant in terms of the cultural and political transitions of post-WWII America. Imagery in paintings by American abstract expressionists Jackson Pollock, Willem de Kooning and others were non-figurative – portraying individual freedom – and purposefully rejected Cubism and American regionalism (Doss 1991). Their unpolished works were seen as a gesture against the glossy commercial advertisements and glitz of commodity worship depicted in popular magazines of the time (Sandler 2009). The "unfinished" look of their paintings has also been attributed to their bohemian and proletariat lifestyles. By the mid-1950s, the Pop Art movement developed in New York by artists Andy Warhol, Roy Lichtenstein and others. These artists were responding to the work of the artists who preceded them, as well as the commercial, consumer-oriented, and popular culture of post-WWII American confidence.

Allen Kaprow invented and coined the term "happening" in 1959, later known as performance art (Kaprow 1966). Artists involved in staging Happenings appropriated the Dadaists' notion of audience participation. Sontag (1966) indicated artists often abused the audience in a non-stage setting congested with objects and by throwing water or other objects. Happenings started in New York and subsequently were staged in Osaka, Japan, Stockholm, Sweden, Cologne, Germany and Milan, Italy (Sontag

1966). Kaprow (1966) defined the happening as an assemblage or collage of events in certain spans of time and in certain spaces. Those engaged in creating Happenings included artists and musicians who were confronting the conventional view of art at the time and wanted to bring art into the realm of everyday life. In parallel with Happenings and Pop Art, "Minimalism" materialized in the late 1960s. Minimalist artists rejected the biographical nature of their work as a unique creation. Their minimalist practices continued through the 20th century with the creation of artifacts that were impersonal and neutral, and had no resemblance to any real objects. Donald Judd, sculptor, and leading minimalist stipulated his works were explorations of space, scale and materials that could be ends in themselves (Kellein 2002).

Protest art, another genre, emerged as civil rights, the free speech movement and American military involvement in Vietnam became intertwined in the 1960s and 1970s. America's post-WWII building boom resulted in environmental degradation by the early 1970s, and spurred the "Land Art" or environmental art movement. Additionally, artists became disillusioned with the commercial and exclusive aspects of having their work exhibited in art galleries. Hence, they sought outdoor locations where their work could be conceptualized and informed by that particular place or environment. Beardsley (1998) carefully articulates these artists' ideology and ways they utilized the land and/or environment as their medium, as well as for their inspiration; and he traces ways artists' temporary and permanent site-specific installations materialized in the public realm.

Modern art over the last 100 years covers a range of genres. Late 19th century and 20th century artists, individually or collectively, were influenced by industrialization and consumer commodification, as well as world politics. Often responding to the turmoil surrounding them and the art world that preceded and constrained them, some artists worked collectively and sought to be the conscience of society. Others sought to lead their society's transformation and shape a new culture and other artists worked individually with little interest in engaging in or shaping society.

Modernism in architecture and landscape architecture

In framing the narrative for modernism in architecture, noted historian Pevsner (1936) drew from the convergence of the Art Nouveau and Arts and Crafts movements in the late 19th and early 20th centuries, along with advances in engineering and technology. According to Pevsner (1936), modernism in architecture temporally followed these movements and defined by Walter Gropius and the Bauhaus School in Germany. Bauhaus thinking in some ways was similar to the Russian Constructivists in that it was rooted in the craft and exploration of materials, as well as unifying the applied arts, fine arts and industrial design. Due to social turmoil and rise of Nazism in Germany, Gropius and others relocated to the USA in the 1930s.

Giedion's (1977) premise for modernism in architecture differed from Pevsner in that his narrative for modern architecture covered architecture and urbanism in the 400 years preceding the 19th and 20th centuries when the emergence of the interplay of art, science and technology were vital in western Europe and the USA. Jencks (1973) expanded the notion of modernism, further noting that one could derive from semiotics, as well as creating classifications for its plurality of approaches. Jencks' narrative for modern architecture factored in socio-political contexts, as well as cultural production outside of architecture.

In broad terms, modernist forms in architecture were stripped of any previous historical references or ornaments, and emphasized structure and materials rather than ornate construction techniques. The rational and efficient use of space was also emphasized. These notions were represented in the term "form follows function", coined by American architect Louis Sullivan, and was considered the modern architect's mantra. Subsequently, Frank Lloyd Wright, an American architect who had worked for Louis Sullivan, modified the phrase to "form and function are one" (Levine 1996). Another pioneering modern architect, Ludwig Mies Van der Rohe (from Germany originally, relocated to the USA due to the rise of Nazi oppression and was based in Chicago, Illinois), quipped "less is more", expressing minimalism as a modern ethic (Mertin 2014). Furthermore, other pioneering modern architects of major note include Le Corbusier (Charles-Édouard Jeanneret from Switzerland originally and based in Paris, France), Richard Neutra (originally from Vienna, Austria and relocated to Los Angeles, California in 1925) and Oscar Neimeyer (Brazilian and based in his native country).

With FDR's New Deal and the 1930s political and economic turmoil in Europe, more thinking emerged in modern architecture by leaders such as Buckminster Fuller, Philip Johnson and Louis Kahn (Frampton 1980). As technology and management systems advanced in the 20th century, so did ideology in modern architecture. Frampton (1980), Jencks (1973), Pevsner (1936) and others have discussed in depth the many important works by modern architects and their followers from different geographies and cultures. By the late 1970s and 80s, Jencks (2002) observed the shift in architecture towards a new paradigm he deemed postmodernism.

Modernism expressed in the field of landscape architecture in the West emerged later than architecture. While some historians claim landscape architecture trends follow architecture with architecture following trends in the arts, the author argues that landscape architecture is its own distinctive discipline with a unique evolutionary path and need not be subordinated to trends in architecture. There are clear inter-relationships among the disciplines of art, architecture and landscape architecture, and sometimes one informs the other, especially with advances in science and technology outside the discipline of landscape architecture. A significant body of literature and resources now exist on modern landscape architecture and its beginnings in the west. This includes numerous built projects (some preserved, some destroyed) by modern landscape architects and their own writings (Eckbo, Kiley, Rose 1939, 1940; Eckbo 1950; Church 1955; Rose 1958; Halprin 1969, 1975; Bye 1981; Walker & Simo 1994; Kiley & Amidon 1999).

Students studying landscape architecture in the USA learn about modernism via the three renegades, Garret Eckbo and his two classmates, Dan Kiley and James Rose, who were studying at Harvard University's Graduate School of Design (GSD). They wrote three essays published in *Architectural Record* (1939, 1940), considered manifestos, and re-published in Treib's (1993) edited book. They were not inspired by the Beaux Arts neo-classical tradition in GSD's landscape architecture curriculum and expressed more interest in Walter Gropius' teaching on the *Bauhaus* approach in GSD's architecture program (Walker & Simo 1994).

Eckbo believed in landscape architecture as an agent for social change. Dan Kiley's subsequent works merged Renaissance-inspired axial geometric forms with modernist thinking – realizing his signature "modern classicist" style. Other important modern landscape architects include Thomas Church, who pioneered modern spatial forms in California residential gardens during the pre-war years and post-WWII building boom. Church's Donnell Garden in Sonoma, California, a modern exemplar completed in 1948, involved the hand of Lawrence Halprin who worked in Church's San Francisco office. Halprin later emerged as an internationally known award-winning modern landscape architect who introduced the community participatory workshop process to the allied planning and design professionals. Roberto Burle Marx, Brazil-based landscape architect, was another major influence on the western world. Other important sources on modern landscape architecture are Tunnard (1938) and The Cultural Landscape Foundation with its accompanying website (tclf.org) established by Charles Birnbaum.

Thematically, this exploration of modernization, modernity and modernism demonstrates the dynamic inter-relationships and convergence of ideas. Modernization involves processes of technological and scientific advances, economic change and their collective impact on socio-cultural processes in society. The western industrial age of the late 19th century, turn of the 20th century, and the march through 20th century modernization theory and modernism in the arts, architecture and landscape architecture, in effect capture transformations in society, as well as intellectual thinking. Modernity in many ways refers to the social and cultural dimensions of these objective changes or the character of everyday life within a new context of changed circumstances. As an experience of adapting to change, modernity existed in an altering, constantly shifting relationship with "modernism" – a cultural reflection or representation and cultural by-product of the modern experience or modern condition.

During the first three quarters of the 20th century in the west, the modern condition and modern society were forever changed by advances in technology, population growth, capitalism, consumerism and worldwide military conflicts. In this shifting context of modernization, modernity and rise of urban life, the artists, literati and intelligentsia expressed a mix of alienation, pessimism, exhilaration, utopianism, collectivism, individualism and activism. Innovations occurred throughout the search for creative expression in the modern western world. At the same time, over the course of the 20th century a dichotomy or tension between artists engaged in representing or interpreting the modern world and participating in its transformation; and those only interested in transforming the cultural artifact itself without interest in social change – art for art's sake, not art as social activism.

Modernity in China

Modernity in China is a vast subject that covers temporal, political, social, cultural and spatial contexts. Unravelling this "vastness" provides a way to understand the public park's emergence in China and late 20th century hybrid modernity. Three distinct narratives contribute to understanding modernity in China – once a largely feudal society transformed from thousands of years of imperial governance guided by Confucianism and dynastic time. The first scholarly narrative argues China's modernity was tied to the arrival of westerners and their influences on China's material culture and Confucian imperial society, so-called westernization and "coerced modernity", or semi-colonialism. A second narrative argues the decline of the Qing dynasty's last imperial court combined with the presence of foreign colonizing powers contributed to domestic disorder. This political context enabled the 1911 Revolution led by the foreign-educated, native-born Chinese, Sun Yat-sen, and caused the demise of the Qing imperial government with the eventual establishment of the modern Republic of

China (ROC). A third narrative suggests a grassroots or homegrown modernity that emerged circa 1915–1919 through the intellectuals of China's "New Culture movement, *Xin Wenhua Yundong*" and the student-led "May Fourth movement, *Wusi Yundong*" (Levenson 1968). This grassroots movement advocated for liberalism, progress, national independence and emancipation from Confucian ideals and the colonizing powers of Europe, United Kingdom, Japan and Russia; faith in western-style science, democracy and individual rights; and more importantly, the strengthening of China in the world context and its construction as a modern nation-state.

China's grassroots New Culture movement was largely responding to the failure of the 1911 Revolution and inability to establish an effective modern Republican government. The push for wholesale reform or cultural renaissance gained more steam with the student-led activism of the May Fourth movement. The students were outraged by the weakness of Chinese government officials at the Paris Peace Conference negotiations for ending WWI and the outcome of the subsequent 1919 Treaty of Versailles (Spence 1991). The treaty created a sense of tremendous betrayal and national humiliation when the Shandong Province, a territory occupied by Germany, was given to Japan. The spirit of these youth movements combined with internal domestic civil strife, warlordism, Japanese occupation and WWII led to the rise of Mao Zedong, the Communist Party of China (CPC) and establishment of the People's Republic of China in 1949.

In the period spanning 1949 and 1978, China has been depicted as a place of agrarian and industrial experimentation, scarcity, suffering, and extremes and contradictions, as Mao attempted to create a socialist China. This period suggests a fourth interpretative narrative, "Mao's Modernity". The following synopsizes these four narratives for modern China. Overlapping themes emerge and provide some context for understanding public parks and greening, and ways these contributed to China's modernity, identity, modernization and nation-building. It sets the stage for Deng Xiaoping's reforms and post-Mao late 20th century hybrid modernization and hybrid modernity.

China's early modernity and westernization

Some scholars argue westernization in China began with the early presence of Christianity (Nestorians and Franciscan Catholic priests) circa 1250s in the Yuan imperial court, ruled by the Mongolians, also known as Mongols and one of the two so-called "foreign" non-Han Chinese emperors. The Manchurians, the other non-Han Chinese, also referred to as Manchus, ruled the later Qing dynasty. The traveler and Italian merchant Marco Polo and his father and uncle, also involved in trade and commerce, were known to have visited the Yuan imperial court. Historians have noted that the two dynasties, Yuan and Qing, ruled by non-Han Chinese, tended to be more tolerant of foreigners and "outward" looking. As mentioned earlier, the Han clan is historically China's dominant ethnic group and intertwined with Chinese identity.

The first papal envoy to China in the Yuan dynasty involved Roman Catholic priests of the Franciscan order. Pope Innocent IV's goal for papal envoys to the Yuan imperial court during medieval times was largely diplomatic. He was seeking military alliances for the so-called religious holy war against Muslims during the European medieval period (De Rachewiltz 1971; Arnold 1999; Bays 2011). The existence and contributions of the Roman Catholic Church to China's Yuan dynasty were less visible than the significant impact of Roman Catholic Jesuit missionaries who graced the courts of the Qing emperors. Conventions in the field of scholarship on the Jesuits in China include the First Jesuit Mission 1552–1773; the Second Mission circa 1842–1949; and a third contemporary period that commenced with the establishment of the People's Republic of China in 1949.

Early modernity and westernization emerged with the Jesuit missionaries and their significant introduction of western knowledge to the Qing imperial court during the 16th and 17th centuries. This included Christianity along with western sciences and culture. It is widely accepted that a robust cultural exchange between western Europe and China occurred as a result of the Jesuit missionaries who served in the Qing imperial courts and immersed themselves in imperial court life. By learning the Chinese language and culture, the Jesuits translated Confucian classics for European intellectual consumption, as well as for intellectual discussions with Chinese scholars in the Qing imperial court. Western and Chinese cultures influenced each other at the time. Many scholars agree that this China/Europe intercultural moment was at its peak as Europeans learned about China, and Chinese learned about western knowledge, science, mathematics and culture. Additionally, it has been generally accepted among art and design historians that the introduction of western perspective and chiaroscuro representational techniques in China occurred as a result of the first Jesuit mission's leader, Father Mateo Ricci (Zou 2011; Keswick 1978).

Some Jesuit priests were also allowed to visit and reside at the Old Summer Palace, *Yuanming Yuan*, located outside Beijing. Also known as the Garden of Perfect Brightness, this vast complex of imperial buildings and gardens was built over 150 years during the 18th and 19th centuries for the Qing imperial court (Wong 2001). Some of the Qing emperors spent more time governing from the Old Summer Palace rather than the imperial complex at the Forbidden City in Beijing. It covered 800 acres and was located five miles northwest of Beijing's walled Imperial City. Its reputation grew to be known as the "Versailles of the East".

Descriptions of the Old Summer Palace with its complex of royal buildings and gardens made their way back to Europe and England and influenced *Chinoiserie*, a late 17th century French term for European interests in Chinese cultural artifacts (architecture, furniture, interiors, furniture, gardening and pottery). This trend began in France and made its way to 18th century England (Jacques 1990). Chinese garden culture and its naturalistic and asymmetrical aesthetics were welcomed by England's garden innovators and writers Alexander Pope, William Kent, William Chambers and Lancelot "Capability" Brown (Hsai 1997; Thomas 2009). Chinese garden influences on English Romanticism and the pastoral landscapes of English parks and gardens were a topic of scholarly discussion in the 18th and 19th centuries (Hsai 1997; Jacques 1990).

Jesuit priests typically brought with them cultural artifacts of western Europe and stories about western royal life to the Qing imperial court. At one point, Emperor Qianlong requested the Jesuits to design and build French, Italian and Austrian palaces with accompanying gardens and water fountains for the Old Summer Palace complex (Zou 2011). Detailed descriptions of this sub-area within the palace grounds known as *Xiyang Lou*, western multi-storied buildings, and this vast imperial garden have been deeply investigated (Wong 2001; Zou 2011). In addition to the design of the western-style garden, Qianlong commissioned a set of twenty copper-plate engravings of the garden views or scenes within this sub-area of the Old Summer Palace and garden complex. Utilization of the copper-plate engraving technique and the representation of the western style garden along with the Old Summer Palace remains a topic of scholarship across various disciplines. Qianlong also commissioned forty paintings of the Old Summer Palace. Copies made of the forty paintings were made using woodblock printing technology and prints of these paintings were disseminated among the literati, or so-called scholar-officials of the imperial court, with an original volume retained by the Qing emperor. Later in 1860 during the Second Opium War, this complex of imperial buildings and gardens was looted and destroyed by a joint effort of the French and English armed forces. The Qing emperor's visual record of the Old Summer Palace was stolen during the looting and is currently held at the Bibliothèque Nationale (French National Library) in Paris (Thomas 2009). Ruins or remnants of the Old Summer Palace grounds have been preserved and currently serve as a tourist site and memorial to the national humiliation of the time.

Forced (colonial) modernity

By the late 18th century and well into the 19th century, China's Qing imperial court was challenged by the British opium trade conducted through the port at Guangzhou (formerly known as Canton), located at the mouth of the Pearl River. Hong Kong and Macau are just south of Guangzhou and frame the gateway to the Pearl River delta. Foreign trade between 1760 and 1842 in China was restricted to Guangzhou and heavily regulated by the Chinese. This so-called Canton System of regulating foreign trade generally enabled three exchanges: local Chinese trade with Southeast Asia; European commercial trade who purchased Chinese goods by carrying merchandise from India and Southeast Asia into China for cash profits; and the commercial trading between Europe and China. The Qing imperial court kept an arm's length from the western "barbarians". Through the Canton System, it designated Chinese merchant companies to serve as middlemen who regularly conducted business directly with the foreign traders. These appointed merchants grew powerful and corrupt and evolved into warlords or supported them financially in the early decades of the 20th century. They contributed to post-imperial China's disunity and internal strife – arresting the development of a modern government system originally intended by the Republic of China (ROC).

The British opium trade caused widespread addiction, social disarray and economic disruption in late 19th century China. Eventually, the Qing imperial court engaged in two military conflicts, the so-called Opium Wars: 1839–1941 and 1856–1860. The first involved the formidable British navy and the second an allied force of both the British and French military. A series of peace agreements or so-called "unequal treaties" were subsequently negotiated. A snapshot of the consequences of the two Opium Wars demonstrates ways that "westernization" was considered a type of forced modernity in China. It also provides the historical context for the moment when the idea of the public park was imported to China for use by foreign residents, not local Chinese. The Opium Wars mark the beginning of China's so-called century of humiliation.

The 1842 Treaty of Nanking (currently Nanjing), an agreement to end the First Opium War, forced the Qing imperial court to open five port cities for foreign trade, as well as cede Hong Kong to British rule. In opening the five so-called treaty ports (Shanghai; Guangzhou; Ningpo, currently Ningbo; Fuchow, currently Fuzhou; and Amoy, currently Xiamen), foreigners (British, American, French, Germans, and others) were allowed to reside and conduct business in designated areas, so-called foreign concession districts where public gardens were created for the foreign expatriate community. Separate municipal bodies, created by foreigners, managed everyday operations of the treaty ports and foreign concession districts. Members of the local Chinese community, generally, were not allowed to reside in the foreign residential districts.

Negotiations to end the second Opium War, circa 1856–1860, demonstrated British and European greed and aggression for *carte blanche* trade and permanent foreign presence in China. A series of unequal treaties were negotiated and included the addition of ten more Chinese ports forced open for foreign trade; legalization of opium trade; British acquisition of Kowloon, land adjacent to Hong Kong Island; the granting of full civil rights to Christian missionaries in China; free navigation along the Yangtze River; and establishing embassies for Britain, France, Russia and the USA in Beijing, formerly known as Peking, in a designated area known as the Peking Legation Quarter (Spence 1991). As noted earlier, the destruction of imperial buildings, including the Old Summer Palace complex of gardens, during the second Opium War contributed to China's national narrative of humiliation. The massive suffering due to opium addiction and semi-colonization by western powers magnified their sense of inferiority to the nations of the world and contributed to the national narrative of suffering and humiliation.

Westernization was often conflated with modernity by the reformers of China's late Qing dynasty imperial court. These reformers, who were desperate to sustain China's system of governance and dynastic rule, led the Self-strengthening movement, *Ziqiang*, circa 1861–1895 – a response to the external foreign influences and Opium Wars. The reformers believed that learning from westerners or the "barbarians" could be a strategy for improving or modernizing their military defense, restoring power and reforming Qing court practices. One of the reformers, Zhang Zhidong, coined the term "Chinese learning for essence, Western learning for function, *zhongxue weiti, xixue weiyong*" and often abbreviated as *Ti-yong*, essence-function (Levenson 1968), a recurring theme to this day.

The Qing reformers' goal was not transformative or progressive, but to somehow "modernize" while preserving Confucian ideology or Chinese essence and thousands of years of dynastic rule. To strengthen itself against the West, the Qing imperial court sought to modernize their military defense, related topics in modern science and western management by learning from the West. This involved the construction of shipyards and arsenals, and hiring foreign advisers to train Chinese for these efforts. It also included sending young Chinese men and government officials to study abroad (USA, England, France, Germany), as well as Japan to learn about the Meiji Restoration and modern government practices.

In addition to the consequences of the Opium Wars, three other civil conflicts are important to point out in the examination of China's modernity: the Taiping Rebellion, a fourteen period (1850–1864) of religious radicalism and a type of civil war between a religious sect and the Qing imperial army; the Sino-Japanese War (1894–1895); and the Boxer Rebellion (1898–1900), an anti-Christian and anti-foreign conflict that began in northern China in the late 1890s and made its way to Beijing in 1900. While dealing with these conflicts, the Qing imperial court experienced waves of other reforms to maintain power in addition to the Self-strengthening movement including: "100 Days of Reform", circa 1898, and "New Policies", also known as the late Qing reform, circa 1901.

Modernization activities in treaty port cities were financed by the western colonial powers. This involved the construction of railway and road systems to facilitate efficient transportation to and from the ports, as well as street lights and public pedestrian walkways, and the "public garden" or "recreation ground" – public parks built for foreign consumption. At the same time, western Christian missionaries were making in-roads in China, establishing churches and schools and converting local Chinese. It is within this context that the fanatical religious quasi-Christian movement, "*Taiping Tianguo*", Heavenly Kingdom of Peace, emerged. Their utopian ideology intended to radically transform, modernize and replace the corrupt Qing imperial government.

The Taiping religion beliefs fused Christianity with Daoism, Confucianism and nativism. Some of the leaders laid claim to having the voice of God or Jesus Christ and saw their military aggression as a type of "holy war" (Spence 1996). They recruited the more impoverished Chinese population into military action against the Qing Empire, captured the southern city of Nanjing – making it the Taiping political and military capital for over ten years. With their inability to capture Shanghai and rally the westerners and missionaries to their cause, along with fifteen embattled years, the Taipings were defeated. The combination of incoherent leadership, failure to gain support from the land-holding Chinese merchants and the effectiveness of the Qing imperial army, bolstered by the British, contributed to the Taipings' defeat. The Christian element of the Taiping conflict left strong biases against Christianity in China (Spence 1996).

Thirty years later, circa 1894–1895, while reconstructing late Qing China from the devastation caused by the Taipings, a military conflict with the Empire of Japan occurred. This involved a territorial dispute over the Korean peninsula. It's important to note that thirty years had transpired since Japan's 1868 Meiji Restoration when their feudal society was transformed and modernized. The Qing's naval and army forces could not match Japan's modern military and were defeated with another unequal treaty for peace signed in 1895. This agreement opened particular Chinese cities to Japan with similar terms previously negotiated with the United Kingdom, USA, France and Russia. This Japanese victory demonstrated the failure of the Qing court's Self-strengthening movement to modernize China's military. This loss also contributed to China's narrative of humiliation and internal suffering.

Soon after the Sino-Japanese conflict, the Boxer Uprising, a peasant-based, anti-foreign and anti-Christian movement, rebellion emerged in north China circa late 1890s and made its way to Beijing in 1900. This group practiced martial arts and was known as the "Righteous and Harmonious Fists, *Yihequan*". Apparently shrouded in mysticism and operating in secrecy, they claimed to have supernatural powers (Spence 1996). Foreigners living in China called them "Boxers" in reference to their martial arts practice.

The Boxers were from the Shandong Province, a northern ecologically fragile area containing the Yellow River and the Grand Canal. People from this area suffered from famine, as well as frequent natural disasters including periods of drought (Esherick 1987). The region's environmental history involved massive flooding of the Yellow River and affected commerce along the Grand Canal, a major economic driver for the area. As an economically depressed area, banditry and lawlessness were commonplace. Shandong's unemployed and impoverished peasantry could not afford the type of education that led to passing civil service examinations, a requirement to work for the imperial government. The Boxers had no interest in Chinese classical education and instead took pride in their martial arts skills and defended those who were victimized by banditry. Disenchanted with the Qing government's weaknesses and lack of resistance to the foreigners, they essentially resented both Germany's possession of Shandong Province and numerous Catholic missionaries. The ongoing drought in Shandong caused additional suffering and it was interpreted as a mystical sign against the Christians and all things foreign (Esherick 1987). Subsequently, the Boxers attacked Christian missionaries, Chinese converts and foreigners throughout the Shandong region and moved south to Beijing by spring of 1900 – destroying foreign-made railways and telegraph lines along the way.

It is important to note the telegraph was a significant part of daily life among foreign residents in Beijing and treaty port cities. The details of the Boxers horrific attacks received extensive media coverage in newspapers around the world and fueled concerns about the safety of the missionaries and foreign expatriates in China. Concerns were magnified when the Qing imperial court decided to support the Boxers. By the summer of 1900, an international military force of 20,000 soldiers representing the "Eight Nation Alliance" (Austria-Hungary, France, Germany, Italy, Japan, Russia, the United Kingdom and the United States) were mobilized to defend Beijing's international district against the combined military forces of the Boxers and Qing government (Esherick 1987). This foreign army occupied Beijing for several weeks, looting and destroying imperial properties including the Old Summer Palace, and terrorizing the Chinese community before announcing their defeat of the Qing army.

The 1901 Boxer Protocol, another unequal treaty, officially ended the three-year Boxer Rebellion and the Qing imperial leaders agreed to a number of activities including erecting monuments in locations where over 200 westerners died; banning all examinations for five years in cities where anti-foreign violence occurred; a two-year period forbidding the import of arms to China; executing the leaders of the Boxers; and payment of 450 million taels as an indemnity for damages and loss of life, around $333 million at the time (Esherick 1987). Later in 1909, the USA established the Boxer Indemnity Scholarship from excess indemnity funds. An estimated 2000 students received Boxer scholarships through a competitive application process between 1909 and 1937, which enabled their studies of various subjects in USA universities. Liang Sicheng, considered the father of Chinese modern architecture and architectural historian, and his wife, Lin Huiyin, were Boxer scholars who studied at the University of Pennsylvania. As early as 1905, Chinese students studied architecture abroad in Europe, Japan and the United States and some studied horticulture in Japan and France (Xing 2002; Lin 2005). Students were also educated by missionary schools in China or were financed by parents to study in Japan, the United States, and Western Europe. After their overseas studies, some of these students returned to lead art academies and university programs in architecture, horticulture and other fields.

Scholars have widely accepted that the Taiping uprising, 1894-5 Sino-Japanese War and consequences of the Boxer Rebellion were emblematic of the Qing court's weaknesses: its lack of capacity to govern, ineffective military and failure to maintain unity. Overall, morale in late 19th century China was low due to natural disasters, economic stress caused by high taxes, and hunger from over-population and the Qing court's inability to keep up with food production. This overall suffering was further compounded by a perceived corruption in the Qing imperial court, trade imbalance, opium imports and massive opium addiction, and the humiliation from the outcomes of the Opium Wars, first Sino-Japanese War, and Boxer Rebellion. The Qing government's decline and weaknesses contributed to its inability to modernize. Control over their major ports was relinquished to foreigners, along with land set aside for foreign-only districts in treaty ports. Some of the Qing reformers were exposed to radical thinking and modern society in their studies abroad, as were other Chinese nationals. This milieu gave way to several revolts, with the most effective involving the 1911 Revolution led by Sun Yat-sen with the eventual establishment of the Republic of China (ROC) in 1912. Most significantly, Sun was then and is now considered a revolutionary hero, and the founding father of modern China.

The public park, Republican modernity (1912–1949) and nation-building

The public park, also known as the "public garden" or "recreation ground", was first introduced in the international or foreign concession districts in China's semi-colonial period in the late Qing dynasty. However, these municipal parks and recreational open spaces in treaty port cities were regulated separately from the Chinese residential districts by foreign municipal officials. These public gardens and leisure spaces were administered as places for consumption by foreign residents, not local Chinese. For example, Shamian Island was created for the French and British residents of Canton, currently Guangzhou, and included so-called public gardens. Similarly, Huangpu Park in Shanghai and parks in other foreign concession districts in treaty port cities excluded local Chinese residents (Skinner 1977; Elvin 1974; Bickers & Wasserstrom 1995).

As noted earlier, park development in Western Europe and North America originated in the 19th century as a response to unhealthy city environments created by the industrialization of cities and towns. This western park-building movement was part of a wider public hygiene perspective that saw parks as a means of improving the health and welfare of the general population, as well as modernizing society. The public park was considered a place where new city dwellers who migrated from rural areas could learn to behave civilly. This movement spurred development of the picturesque parks in England, Olmsted's Central Park in New York, and a subsequent park-building movement throughout the United States. Following the French Revolution, royal gardens throughout France were transformed into public parks as a demonstration for civil liberty. The later renovation and modernization of Paris circa 1852–1870 and Haussmann's plan for a city-wide system of public open spaces as the "lungs of Paris" demonstrated the relationship between public health and civic design. The plan called for major public parks at the four major cardinal locations in Paris, renovations of French royal parks, tree-lined boulevards and pedestrian promenades connecting parks and squares.

This social hygiene movement had an indirect effect on the development of parks in China. While Shanghai's Municipal Council, a managerial organization comprised of foreigners who managed the International Settlement, did designate the "Chinese Public Garden" for local Chinese use in 1890, it would be several years before municipal leaders in modern China established public parks for the local Chinese community (Bickers & Wasserstrom 1995). Public parks in China's treaty port cities were created for foreign consumption only.

Limited access to the parks in Shanghai's International Settlement was known throughout China as exclusionary and discriminatory toward local Chinese. Controversy surrounding text in signage at entries to the foreign concession parks,

"No dogs or Chinese allowed", contributed to nationalist anger, suffering and humiliation (Bickers & Wasserstrom 1995; Jackson 2017). Shanghai's public gardens in the International Settlement were symbols of semi-colonialism and "Old China" imperialism, and social inequality represented in the treaty port world (Jackson 2017).

Modern China's first designated public park for local use occurred in Beijing, China's long-standing capital city, circa 1914. It was created through the adaptive re-use of an existing imperial garden located outside the southern city wall of the Forbidden City and across from present day Tiananmen Square. Beijing's Central Park, now known as Zhongshan Park, was built as part of nation-building and modernization in the early years of the ROC.

Shi (1998) notes the modern idea of Central Park was imported to China from the west via Ueno Park, Tokyo. Beijing city officials were inspired by the American experience of municipal administration and established the Municipal Council of Beijing. This council was responsible for modernizing the city's physical environment with public works programs including building public parks for local use – a foreign idea that was essentially unknown in China until this period. Shi (1998) noted Beijing officials' site selection criteria for Central Park: located in the center of the city; on the site of a former imperial garden; and within the walls of the Forbidden City (Figure 2.1). Beijing's goal was to demonstrate to the world and international community that Beijing was a modern city.

Beijing municipal government officials were strategic and financially practical when selecting the former Altar of Earth and Grain, a sacred ceremonial site for Central Park.

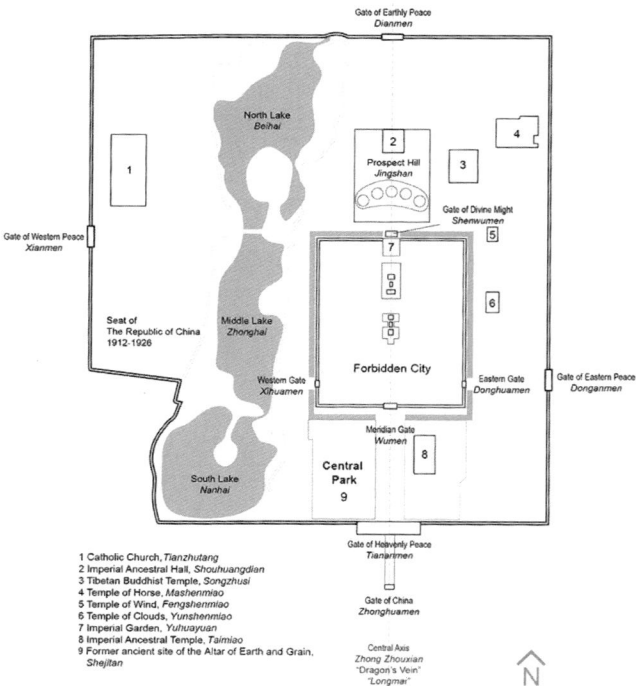

Figure 2.1 Central Park location in Republican era Beijing, graphic by X. Liu

Figure 2.2 Ancient tree in Beijing's Central Park, renamed Zhongshan Park in 1926, photograph
by author

Transforming a former imperial garden meant the government avoided significant
costs associated with the construction of a new purpose-built park. Mature stands of
cypress trees (Figure 2.2) planted before the Ming dynasty and existing buildings could
readily be incorporated into Beijing's Central Park. New park improvements included
a path system, rockery and artificial hills typically found in traditional Chinese gardens, as well as remnants from the Old Summer Palace (X. Sun 2008, pers. comm., 18
June). Several existing halls were converted for use as educational exposition buildings
and other government use. The park became a leisure space where Beijing city residents could socialize and enjoy the natural scenery (Shi 1998).

Beijing's Central Park was shaped by an interesting combination of foreign and
Chinese historical influences. In addition to its connections to major parks in New
York City and Tokyo, there were clear connections between Beijing's Central Park
and its imperial past. The park contained modern artifacts from the imperial and
ROC eras, and rockery and artificial hills typically found in traditional Chinese gardens. Several Qing-style buildings were converted for administrative use, a restaurant, and a museum dedicated to Sun Yat-Sen. While the design of Beijing's Central
Park was not innovative, its creation represented a shift toward the formal establishment of city public parks in modern China intended for use by local residents. In
some ways, it was not unlike the conversion of royal parks in Europe and England
for public use.

At the time, heavy industry did not yet exist in Beijing, and the urban pollution of industrial cities like London and New York had not yet developed in China. It was transforming from primarily agrarian and feudal life into a modern society. Parks created during this period in China were part of a program of modernization that brought public infrastructure improvements such as road construction and street systems, railroads, pavement and public walkways, and street lighting to Chinese cities that were not necessarily treaty port cities (Esherick 2000; Dong 2000; Wasserstrom 2000). In Hangzhou, Zhejiang Province, for example, the ancient city walls were demolished, a new city grid was imposed and a series of waterfront public parks were built along West Lake. The movement toward parks development during this period was driven more by the desire to modernize along European lines than by the type of social concerns about hygiene that drove the development of Central Park in New York. Beijing's Central Park, *zhongyang gongyuan*, was intended to demonstrate to the world that Beijing was a cosmopolitan city on par with London, Paris, and Tokyo (Shi 1998). In effect, China's first public park in Beijing, the ROC's earlier capital city, was a symbol of modernity to the world.

In the early years of the ROC, several shifts occurred as the nation attempted to remake its everyday feudal life and transform into a modern society. Western practices were introduced. For example, the Edwardian calendar was adopted; the modern work week introduced; and the Confucian-based civil service examinations were replaced with western-style education. Throughout China, major cities were engaged in modernizing. Old city walls were demolished; grid-iron street systems introduced; government construction of railways and public infrastructure systems occurred; and public parks were designated in central city districts. Like Beijing, other cities in China created new public parks through the adaptive re-use or re-purposing of imperial outdoor grounds and sacred places, and named them Central Park. These park-making activities were seen as emblematic of Sun's Three Principles, *Sanmin Zhuyi*, for building a modern nation. Sun's three principles: nationalism, democracy and livelihood were equated to the USA's notion of "government of the people, by the people and for the people" (Lee 1930). The newly designated public parks were considered part of nation-building and symbols of a democratic and modern society. Public parks were places where modern Chinese citizens could relax, socialize and spend their leisure time.

The introduction of new public parks, a "modern" city-making move, was a major transformation from China's traditional city-making principles. The north-south orientation, square lay-out of China's ancient and imperial cities, system of gates, and spatial locations of the principal structural elements (south-facing royal palace, imperial court, ancestral hall, Altar to the God of the Earth, and market) were prescribed by the *Confucian Rites of Zhou* or Book of Zhou, *Zhouli*, and later Record of Trades, *Kao Gong Ji* (Wright 1977; Xu 2000). Areas designated for public parks were not part of the ancient or classical Chinese city cosmology. Activities in Republican era public parks somewhat replaced the leisure, social and commercial activities that took place at temple fairs during religious holidays.

It's important to reiterate the 1912 establishment of the ROC was short-lived. While Sun Yat-sen's 1911 Revolution did topple the Qing imperial court, it was difficult to establish a modern government and unify the nation. By 1916 or so, some political and military factions left over from the Qing's regionally organized army caused the Republican nation to devolve into the so-called warlord era, a period when regional autonomy and territorial conflicts created disunity until the late 1920s (Spence 991). During China's warlord era, Beijing continued to operate as the ROC capital city where world leaders conducted diplomatic affairs.

However, a surge of national hope took place after Sun Yat-sen's death in 1925 during the so-called "Nanking decade" (Wade-Giles spelling of Nanjing) circa 1926–1937 when the Nationalist Party, *Guomindang*, and their army were able to control the warlords and unify China as a modern republic. Then, major efforts were undertaken to honor Sun's original intentions. This involved the commemoration of Sun Yat-sen's death with the name of Beijing's Central Park, changed to Zhongshan Park in 1926, and reflected the *Putonghua* translation of his Cantonese name, Yat-sen. Throughout China, other cities would follow suit and rename their parks to honor the revolutionary hero who led the downfall of the Qing imperial government. Numerous Sun Yat-sen or Zhongshan memorials and monuments were built in public parks and city squares throughout China. This hero worship renewed China's optimism and Zhongshan became part of the lexicon for modern nation-building under a unified government led by the Nationalist Party. Another major nation-building effort that commemorated Sun Yat-sen was the establishment of the National Tree Planting Day on the anniversary of his death, 12th of March. This annual date for national tree-planting practice continued during Mao's tenure, with the name later changed to National Arbor Day in 1979.

The Nanking Decade focused on rebuilding the southern city of Nanjing as the ROC's capital city. With Nanjing as the capital of modern China, China's leaders understood the tremendous responsibilities that went with its urban transformation into a new capital city and the need for it to reflect a new political order. In this light, the model capital for modern China was conceived as a sacred space where modern Chinese citizens would be transformed; and it would serve as a symbol of nationalism – a modern metropolis with Chinese essence (Musgrove 2013). Soon after Sun's death, a committee was established for the new Capital Plan, *shoudu jihua*, and included Sun Yat-sen's son and world traveler, Sun Ke and the American-educated engineer, Lin Yimin, and Mayor Liu Chi-wen, a British-trained city planner (Wagner 2011). They retained American architect Henry K. Murphy as the chief architectural advisor for the Nanjing Capital Plan from 1927–1930. The committee was apparently inspired by western modernization and city-making practices in London, Haussmann's Paris, Vienna, Austria, and especially Washington D.C.'s capital planning and America's City Beautiful movement.

The committee's vision imagined tree-lined boulevards and street corridors that would link the various public urban gardens and squares, and create an overall image of the city as a big garden for public use. The committee prioritized the incorporation of public gardens, greening and open spaces along the lines of American and European concepts – where tree-lined boulevards would emphasize the civic qualities and sense of procession through a renewed Nanjing with the main arterial road, Sun Yat-Sen Boulevard, *Zhongshanlu* (Cody 2011). In some ways, the committee was honoring Sun Yat-sen's belief in "greening", "verdurization", *luhua*, or tree planting as a strategy for building a modern nation (Lu 2018).

It is important to note that Sun conducted in-depth studies of the French, American and Russian Revolutions at the British Library during a nine-month stay in London circa 1896. His studies also led him to understand the symbolism of trees as part of the lexicon of political freedom. The so-called liberty tree was an American symbol of freedom from the British colonizing forces circa 1765 and in France, *arbre de la liberte* also symbolized its freedom from the monarchy (Schlesinger 1952; Harden 1995). He experienced St. James and Hyde Parks, royal gardens transformed into public parks during his 1896 studies in London, and also toured Haussmann's Paris and other European cities while seeking financial and political support for his efforts to overthrow the Qing imperial government.

The 1929 Capital Plan designated Purple Mountain, *Ziji Shan*, part of a mountain range in eastern Nanjing, a national public park. In China, pilgrimages by the general population to sacred sites, like the Ming tomb on Purple Mountain or Buddhist caves near West Lake, Hangzhou, were customary during religious and national holidays. Chinese also visited Zhongshan Parks in various cities for a small admission fee. However, the concept of a public national park had not yet been established. This designation of Purple Mountain served as a significant act of modernity and nation-building from the perspective of national-level planning and park-making in 20th century China. Several decades would pass before China's State Council officially designated forty-four national parks, *guojia gongyuan*, in 1982 (Padua 2014). This official list included Nanjing Zhongshan National Park, which contains Purple Mountain. In post-Mao China, the accepted nomenclature or classification that is the equivalent to American national parks is "National-level Scenic and Historic Interest Area, *Guojiaji Fengjing Mingshengqu*".

Nanjing's Purple Mountain in China's Republican era is a significant precedent for the planning and design of public open spaces, and preservation of cultural landscapes. The 1929 Capital Plan spatially locates this newly zoned national park adjacent to the city districts designated for government administrative center and commercial business district. It contains the site of the Ming ancestral tomb or mausoleum completed in 1401 for Emperor Hongwu, the founding emperor of the Ming dynasty whose capital was the site of Nanjing. Purple Mountain is also the site of Sun Yat-sen's Mausoleum complex, a project that was a result of an international design competition held in 1925 and constructed in two stages: 1) the buildings and designed landscapes for the Mausoleum were completed in 1928 by the Cornell University educated architect, Lü Yanzhi (1894–1929), whose competition entry was selected from over forty entries (Lai 2005; Wagner 2011); and 2) outdoor amphitheater and music stage completed in 1933 and designed by Yang Tingbao (1901–1982), an architect educated at the University of Pennsylvania in the Beaux-Arts tradition (Xue 2008; Ruan 2011).

The Capital Plan committee oversaw an international design competition, the first of its kind in China (Lai 2005). Their objective was to honor Sun Yat-sen's deathbed request to have his remains interred at Nanjing's Purple Mountain. The design competition called for a master plan with buildings in a classical Chinese style with distinctive monumental features, and a memorial hall, tomb and assembly area for 50,000 people, a public square and a musical pavilion and stage. The Sun Yat-sen mausoleum project triggered the beginning of a national architectural style and contributed to the transformation of Nanjing as the nation's modern capital (Andrews & Shen 2012; Wagner 2011).

Lü's winning entry (Figure 2.3) consisted of an axial design, processional path with an entry plaza and gate, ascending steps and landings, and mausoleum structure within a bell-shaped plan, enhanced by tree plantings (Figure 2.4). The spatial form and experiential design was inspired by Chinese sacred architecture, Philadelphia's Liberty Bell, and Sun's three principles (Lai 2005; Wagner 2011). The committee found Lü's

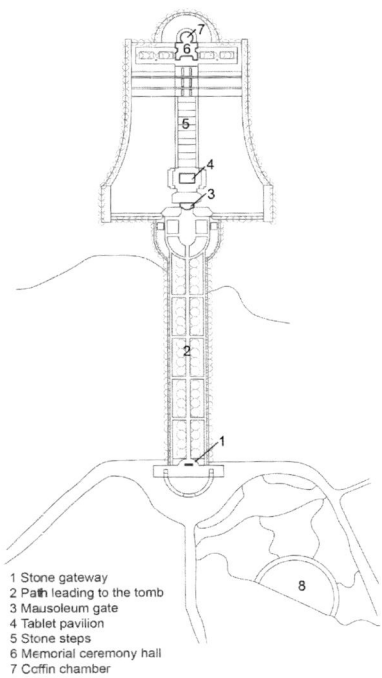

1 Stone gateway
2 Path leading to the tomb
3 Mausoleum gate
4 Tablet pavilion
5 Stone steps
6 Memorial ceremony hall
7 Coffin chamber
8 Music stage

Figure 2.3 Sun Yat-sen Memorial Plan based on Lü's 1925 winning entry (Wagner 2011), graphic by X. Liu

Figure 2.4 Axial processional path aligned with trees, photograph by author

incorporation of the landscape, including tree plantings along the grand staircase and ceremonial path, into his design significant. No other entries considered a holistic master plan that integrated the landscape (Wagner 2011).

The broader plan and built works for Nanjing's Purple Mountain demonstrates the emergence of modern landscape architecture in China during the Nanking Decade. Purple Mountain was designated a national park, and contained China's first botanical garden, known as Zhongshan Botanical Garden. It originally housed a variety of plants given by governments and organizations from around the world in honor of Sun Yat-sen. Another project of note in Lü's winning Sun Yat-sen plan is the outdoor "grand" music stage, *yinyue tai*, completed in 1933. Lü's sudden death after the Sun Yat-sen memorial buildings were completed provided the opportunity for Yang Tingbao, one of China's noted modern architects who studied architecture at the University of Pennsylvania around the same time as Louis Kahn (Cody 2011). Yang finalized Lü's idea for the music stage (Ruan 2011).

Yang's implementation of Lü's plan incorporated the design of the outdoor amphitheater that combined western, neo-classical Beaux Arts style with Chinese traditional architecture. The outdoor amphitheater's semi-circular form, modelled on the Greek amphitheater, incorporated a terraced lawn (Figures 2.5 and 2.6) for open seating (on the ground), and was considered novel for the time (Ruan 2011). Along the perimeter of the amphitheater's upper outer edge, the semi-circular form incorporated a pedestrian path with benches under a vine-covered pergola and colonnade (Figure 2.7). The design of the colonnade's columns was inspired by Buddhist temple architecture and

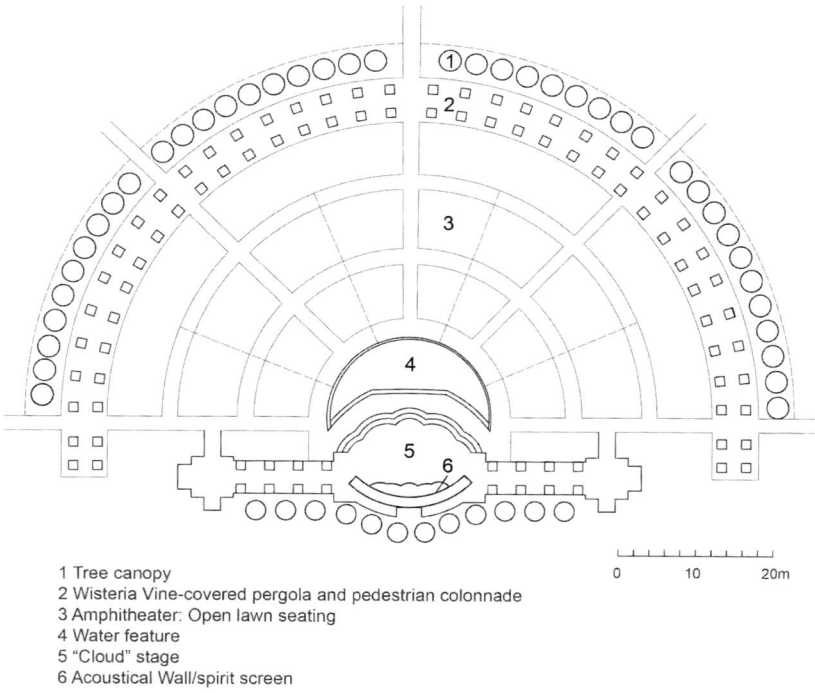

1 Tree canopy
2 Wisteria Vine-covered pergola and pedestrian colonnade
3 Amphitheater: Open lawn seating
4 Water feature
5 "Cloud" stage
6 Acoustical Wall/spirit screen

0   10   20m

Figure 2.5 Yang's plan circa 1930s, based on drawings in Xue (2008), graphic by X. Liu

Figure 2.6 Terraced lawn amphitheater with crescent moon-shaped water feature fronting the stage, photograph by author

Figure 2.7 Vine-covered pergola with columns inspired by Buddhist temple architecture, photograph by author

constructed using reinforced concrete (Xue 2008). A line of trees form the outer edge of the pedestrian-oriented colonnade. At the low point of the amphitheater's sloping lawn, Yang's design incorporated a crescent moon-shaped water feature in front of the raised performance stage. The water feature was designed to serve several purposes: collect stormwater from the grounds of the amphitheater; habitat for fish and lotus plants; and as an ornamental water fountain (Ruan 2011). The design of the stage (cloud shape in plan form) and the curved screen wall, an acoustical backdrop, was inspired by Buddhist temple architecture (Xue 2008; Ruan 2011).

Landscape design education emerged during China's Republican era and grew out of horticultural and forestry programs, and schools of architecture. Early educators were schooled mainly in France and some in Japan (Lin 2005). The first program of horticulture education was offered in 1912 at the Jiangsu Province Second Agriculture School, formerly known as Suzhou Agricultural Vocational Technical College (Lin 2005; Yu 2007). Later, from 1922 to 1927, courtyard design was offered as a separate course at the same school and focused on residential design (Lin 2005). Suzhou was both an important commercial area and a place known for high culture; and it was also the home to hundreds of classical Chinese gardens that still existed during the Republican period. As ROC's capital, Nanjing became an important place for educational institutions (Lin 2005). The influence of foreign ideas and modern design approaches was present everywhere in Nanjing's new educational institutions of the Republican era (Z. Bao 2009, pers. comm., 29 September). For example, the Forestry Department offered garden design courses and was established circa 1931 at National Central University, now known as Southeast University in Nanjing. Southeast University had also established China's second architecture school in 1928; and it offered courtyard design as part of its curriculum (Z. Bao 2009, pers. comm., 30 April).

In summary, during China's Republican era, the idea of the public park was imported from the west, "outward" looking, and symbolized international modernity. Public parks and greening were part of a strategy for nation-building throughout cities in China. These transformed imperial gardens and sacred grounds, new street tree plantings and public parks were part of modernization efforts throughout cities in China. Sun Yat-sen's death in 1925 and the Nationalist Party's last ditch effort to re-unify China during the Nanking Decade created a hopeful time when public parks in cities throughout China became places to commemorate the revolutionary hero who spearheaded the downfall of the monarchy and the creation of a modern nation. Zhongshan's name was affiliated with public park-building and became part of the lexicon for nation-building throughout modern China. Nanjing's Purple Mountain emerged as a place where the built environment, especially the designed, cultural and natural landscapes, was intertwined with nation-building. The argument can also be made that Beijing's initial efforts to create Central Park and the Nanking Decade's park-building and greening efforts were also responding to the national humiliation created by the discrimination towards local Chinese in foreign concession parks in Shanghai. New modern institutions were established throughout cities in China during this decade. More critically, cultural awakening occurred throughout the nation as restrictions in treaty port cities were relaxed and students returned from studying abroad. In parallel, foreign trade activities in the treaty port cities continued to operate.

From a western perspective of that time, China's Nanjing Decade may not have represented grand city-making, architecture or park-making innovation. However, street tree installations were seen as highly effective (Cody 2002; Musgrove 2013). Public parks and tree plantings in the Republican era celebrated liberty and were symbolic of Sun's notion for a modern nation and democratic society. In the post-Mao era, Sun's legacy was honored by retaining the date of the anniversary of his death for National Arbor Day, as well as the adoption of the Fifth National People's Congress circa 1981 of the "Resolution on the unfolding of a nationwide voluntary tree-planting campaign"; and called for able-bodied citizens (age eleven through sixty years) to plant three to five trees per year or equivalent (Richardson 1990).

China's short period of unification during the Nanking Decade was disrupted by the 1937 aggressive invasion by the Japanese who occupied China until the end of WWII circa 1945. During this period, China's natural landscapes, especially trees and forests, were cannibalized by Japanese semi-colonization activities (Edmonds 1998). Between the end of WWII and 1949, China's built and natural environments were neglected with areas destroyed during the domestic guerilla warfare involving militia from the Nationalist (Guomindang) and Communist Parties. The short-lived ROC was replaced after Chairman Mao Zedong's successful peasant-based communist revolution and the establishment of the People's Republic of China in October 1949.

Mao's modernity: 1949–1976

The physical, social and economic environment Mao inherited in October 1949 was ravaged by domestic military and civil strife, Japanese occupation and fractured periods of inconsistent governance and disunity in the Republican era. Furthermore, much of China's natural environment was seriously degraded and re-shaped during the Qing dynasty. This was a result of imperial court needs (agricultural cultivation, hydraulic works for drainage, irrigation and flood control, and deforestation for construction), reconstruction after the Taiping Rebellion and late 19th century modernizing efforts by western colonizing powers (Smil 1984; Elvin 1998). The attempts to modernize and build a nation in the Republican era eventually gave way to Mao's emphasis on industrialization and collectivization of the economy.

Mao's period has been characterized as a period of radicalism, scarcity, experimentation and "contradictions and extremes". Many scholars have focused on the "darkness" and policy failures of Mao's last ten years of governance, a period that included famine and related deaths, the Cultural Revolution and global isolation (Lewis 1963; Smil 1984). Mao believed in accomplishing three goals for China: 1) national unity; 2) social and economic change; and 3) freedom from foreign interference (Westoby 1979). Mao had the task of remaking China around the CPC principles for a unified socialist nation. He chose Beijing as New China's socialist capital city.

With the incredible task to rebuild China, no financial resources and no real experience, Mao sought aid from the Soviet Union, formally known as the Union of Soviet Socialist Republics (USSR), the only other communist state in the world and the first of its kind (Bray 2005). Eventually, the 1950 Sino-Soviet Treaty of Friendship, Alliance and Mutual Assistance, *Zhong-Su Youhao Tongmeng Huzhu Tiaoyue*, often abbreviated as the Sino-Soviet Treaty of Friendship, was negotiated (Spence 1991). It was varied in its scope and involved financial loans to China with high interest rates; 20,000 Soviet experts paid for by China; 80,000 Chinese students allowed to learn science and technology in the USSR; China's ceding of the two ports, Dalian and Lushun in the area previously known as Manchuria (the current northeastern provinces of Heilongjian, Jiling and Liaoning); and China's ceding of mineral rights in Xinjiang

(Spence 1991; Bray 2005; Lüthi 2008). This was accompanied by deep distrust of their former western allies in World War II. Political ties with the Soviet Union became contentious by the late 1950s and dissolved by 1960. At the time, all of China's institutions were transformed to follow the Soviet model.

Soviet technical advisors introduced a socialist-oriented standardization of architecture that was reflected in factories and housing (Rowe & Seng 2002). Although park construction and innovations in park design was minimal early on, Mao did embark on "greening" and afforestation efforts as part of a program of production-oriented landscape development. A slogan known as "Making Green the Motherland" or "Our motherland is a flowering garden", *Luhua zuguo*, was widely used in 1956. It was drawn from the National Program for Agricultural Development launched in 1956 (Westoby 1979; Zhao & Woudstra 2012; Lu 2018). This program was largely perceived as a nationwide afforestation effort to mobilize the masses through the planting of trees or greening near houses, in villages, along roads, as well as along river corridors; and these efforts had its own slogan, "Four-side Greening, *Sipang Luhua*" (Zhao & Woudstra 2012). Similar to Soviet ideology, "greening" was perceived as a metric for communism and establishing a socialist paradise. China's central government sponsored roadside tree-planting activities, renovation of parks and new park construction with additional goals for beautification, *meihua*, and making Chinese cities garden-like, *huayuan chengshi* (Westoby 1979; Lu 2018). However, due to the government's lack of organizational and administrative skills, and a shortage of trees, projects in this greening program were poorly planned or never implemented in the long run (Lu 2018).

It's important to note that tree plantings along roads have been a practice in China since the Qin dynasty when a system of tree-lined imperial roads was developed for use by the emperor and government officials. Like ancient Rome and other civilizations, China's imperial roads were a form of communication, and developed into a type of imperial postal road. The design of the imperial road was prescribed by the Book of Zhou, *Zhouli*, with trees closely spaced to create shade and comfort for the emperor (Yu et al. 2006). The tree canopy optimized shade and visual effects, and functioned as a form of orientation and a way to mark the road. This allowed government couriers (on foot or horseback) to find their way during inclement weather, e.g. snowy winter, blizzards and sandstorms (Profous 1992). In the Mao era, activities related to urban public parks dealt with maintaining existing parks from the Republican period with some new parks based on the Soviet Union model "parks of culture and rest" (X. Sun 2008, pers. comm., 18 June).

Parks of culture and rest, also known as parks of culture and leisure, are considered the Soviet Union's contribution to the lexicon of modern park design. The Soviet Union's socialist intention for this park type was to provide a large park with a diversity of cultural enlightening activities, entertainment, organized sports, education, a place for relaxation and rest in an optimal natural environment (Bittner 1998; Hayden 2005). Gorky Park (initially called Central Park) in Moscow was the Soviet Union's first park of culture and rest built in 1928. It included a variety of design elements that were spatially organized around function or activity zones, not aesthetics, and included a sports stadium; a large square with a road to accommodate military tanks; buildings (theater, music performance, science and technology, library, movie cinema); children's area complete with supervised care; a big Ferris wheel; carousel; restaurants and cafeterias; boating lake and separate location for swimming; quiet sitting areas in natural settings and forested pedestrian paths; and areas for flower displays (Hayden 2005). Soviet parks of culture and rest were also a place for the state to propagate official political doctrines. For example, banners of Stalin's popular slogan, "Life has become better, life has become more cheerful" were hung at the entry gates to Gorky Park (Hung 2013). At that time, the Soviet Union saw "greening" (tree plantings along streets, construction of public gardens, parks and squares) as a way to remake Moscow as their nation's capital city and build a new socialist society (Bittner 1998).

Public parks and greening in the Mao era mimicked the Soviet model. Also known as socialist parks, *shehuizhuyi gongyuan*, the intention was that these public open spaces should be different from the past private gardens – serving the politics of the proletariat. The so-called proletariat in the Mao era was a broad term that initially drew from the Marxist notion of "working people" and subsequently expanded to include industrial workers, peasants, anyone who performed manual labor and the military (Hung 2013). Public parks in China's Mao era were also known as people's parks, *renmin gongyuan* – strategically designated and state controlled public spaces for disseminating political propaganda and providing cultural education to the general public. Generally, parks in China charged a small entry fee, were enclosed with fencing and their gates locked after dark to disallow access.

Early in Mao's tenure, China's mood was celebratory and hopeful. A building boom occurred in the first four to five years as the CPC stabilized the nation. This involved the creation of the square at *Tiananmen*, Gate of Heavenly Peace, in the early 1950s, originally the major southern entry to Beijing's Imperial City and along the north-south axis (Hung 1991). Subsequent to Mao's Making Green the Motherland campaign, the Great Leap Forward (GLF), a modernization program was launched in 1958. Mao's objective was to create an advanced socialist nation through industrialization and agriculture, along China's own independent trajectory, away from Soviet influence. Simultaneously, the CPC announced their intentions to celebrate socialist China's 10th inaugural anniversary with the design and construction of Ten Monumental Buildings, *Shi Da Jianzhu*, including the expansion of Tiananmen Square to accommodate 600,000, for completion by 1 October 1959, or less than ten months. As Mao attempted to unify and create a socialist nation, his period of leadership was dominated by experimentation and standardization through wholesale adoption of the Soviet model. An array of policies, the launching of campaigns, sloganeering and political propaganda were also typical of Mao's government practices, and to some extent carries on today.

Mao's staunch views were largely drawn from Marxist-Leninist theories of socialism. It has been generally accepted that Mao had an "anti-city" or anti-urban perspective. This was due, in part, to his belief that cities, particularly the treaty port cities, were capitalist artifacts reflecting elitist thinking of both the Republican era and feudal society of the earlier dynastic rule. He also believed that Confucianism and traditional Chinese values were linked to city-dwellers. This convergent perspective led to the strong anti-urban sentiment during the Mao era.

Mao believed his political support and the success of his communist revolution was a result of the rural peasantry. Hence, the emergence of the CPC's party ideology and Mao's focus on the "struggle of the Proletariat". However, as part of state control over its citizens, Mao established the official household registration system, *hukou*, a type of domestic passport that established one's identity, citizenship and proof of official status. This system created a bifurcated society and made distinctions between urban and rural dwellers. The state allocated resources and comprehensive entitlements for city residents: e.g. food, water, education, housing, police protection, employment, retirement benefits and free or subsidized health care (Cheng & Selden 1994). None of these entitlements, so-called "iron rice bowl, *tie fan wan*", were allocated to those born in rural areas. However, these types of resources may have been allocated by collectives in rural areas. This household registration system represents the contradictory nature of Mao's policies. On the one hand, he expressed defiance towards city dwellers as elite and bourgeois, yet gave them comprehensive entitlements. On the other hand, his so-called rural peasant base weren't entitled or allocated comparable compensation from the central government or so-called "state".

Before delving into a discussion of other campaigns in the Mao era that may have affected the natural environment, discipline of landscape architecture or innovation in modern park design, a noteworthy new public park, not representative of the Soviet park of culture and rest model, was completed in 1954. The park is known as Viewing Fish at Flower Harbor Park, *Huagang Guanyu Gongyuan*, and located in Hangzhou, Zhejiang Province along the waterfront at the original and iconic West Lake, *Xi Hu* (Figure 2.8). The project was designed by Professor Sun Xiaoxiang (1921–2018) circa 1951–1952, China's first modern landscape architect who was

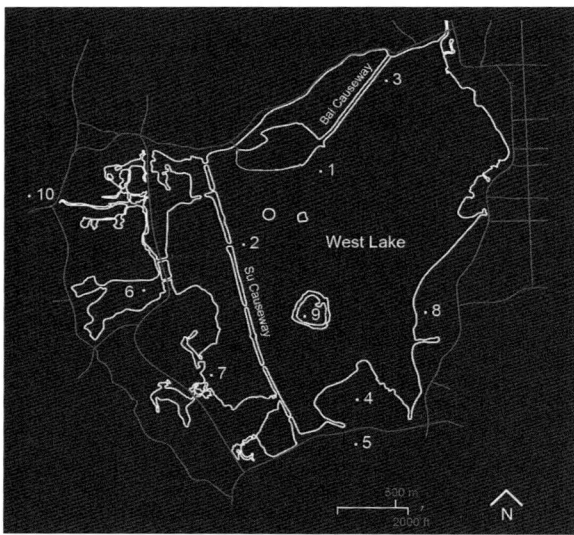

Ten (poetically named) Scenes, West Lake, *xihu shijing*

1 Autumn Moon over the Calm Lake, *pinghu qiuyue* 平湖秋月
2 Spring Dawn Breaking Over Su Causeway, *sudi chunxiao* 蘇堤春曉
3 Lingering Snow on the Broken Bridge, *duanqiao canxue* 斷橋殘雪
4 Leifung Pagoda in Evening Glow, *leifeng xizhao* 雷峰夕照
5 Evening Bell Ringing at Nanping Hill, *nanping wanzhong* 南屏晚鐘
6 Breeze-ruffled Lotus at Quyuan Garden, *quyuan fenghe* 麯院風荷
7 **Viewing Fish at Flower Harbor, *huagang guanyu* 花港觀魚**
   site of Professor Sun's park circa 1954
8 Orioles Singing in the Willows, *liulang wen ying* 柳浪聞鶯
9 Moon Reflected on Three Ponds, *santan yinyue* 三潭印月
10 Twin Peaks Piercing Clouds, *liangfeng chayun* 兩峰插雲

Figure 2.8 Map of West Lake's Ten Scenes and Sun's Park, Hangzhou, Zhejiang Province, graphic by J. Wu

based at Beijing Forestry University. The following discussion of Sun's lakefront park in Hangzhou is based on an interview with Sun at his Beijing residence on the 18th of June in 2008.

Sun's undergraduate education took place at Zhejiang University's Department of Horticulture circa 1940s, and for a year he pursued graduate studies at Southeast University with noted architectural historian Liu Dunzhen (1897–1968). One important note is Sun was a longstanding educator at Beijing Forestry University's landscape architecture program who often represented China internationally, especially in the 1980s soon after China opened to the world. He taught several generations of landscape architects in China including Yu Kongjian, current Dean of Peking University's College of Architecture and Landscape Architecture, and President of the award-winning practice Turenscape. Yu is currently a major voice in China and internationally. Yu is a staunch CPC member; whereas Sun disclosed that he was never a CPC member and believed design should transcend the political context.

Sun's park, Viewing Fish at Flower Harbor Park, *Huagang Guanyu Gongyuan*, incorporated one of the "Ten Scenes of West Lake", *Xihu Shi Jing*, a series of lakefront gardens circa late 13th century, and China's first known system of open spaces immortalized by painters, calligraphers and poets for thousands of years (Sun 1994). These poetically named scenes were at the peak of their popularity during the Song dynasty when Hangzhou was the temporary capital city of China. Later, Qing Emperor Kangxi circa 1699, revitalized the cultural significance of West Lake's Ten Scenes and began inscribing four-character poetic epithets on stone tablets at the site of each of the ten scenes, and his grandson Emperor Qianlong later inscribed his own poetry at the same locations (Barmé 2012). According to Sun, his project at West Lake's waterfront was the first park in Mao's new China to incorporate an open lawn area, a spatial form representative of the English Picturesque garden design vocabulary.

Coincidentally, in 1954, around the time Professor Sun's park was completed, Mao began spending substantial time in Hangzhou, Zhejiang Province, near West Lake. Mao started writing the Chinese Constitution there; and through 1960, he made it a practice to reside in Hangzhou for months at a time (Barmé 2012). Lakefront villas with sequestered gardens in West Lake, Hangzhou, functioned periodically as Mao's base for CPC operations. Mao's time at Hangzhou's West Lake would also involve mountain hiking, calligraphy-making and poetry writing. CPC officials were concerned that Mao's political retreats to luxurious waterfront villas in Hangzhou would go against the grain of frugality and perceptions of an impoverished party and nation. Hence, Mao's time spent at West Lake was explained away by the vague term "the necessities of work", *gongzuo xuyao* (Barmé 2012).

Professor Sun described himself as a traditional Chinese scholar with literacy in poetry, calligraphy, and landscape painting during the 2008 interview at his Beijing residence. His design approach was generally based on the importance of Chinese "garden art" and the spatial ordering through a series of garden scenes, as well as the "garden within a garden" concept. The next chapter will discuss further the language and vocabulary of traditional Chinese gardens. Regarding his design approach for Viewing Fish at Flower Harbor Park, Sun noted the mayor of Hangzhou had originally requested a Soviet culture and rest park. During his research for the park's design, Sun discovered that parks of culture and rest were planned in other areas near or fronting West Lake (Figure 2.9). He successfully persuaded the mayor of the significance to create a new interpretation of a modern park that didn't follow the Soviet standards.

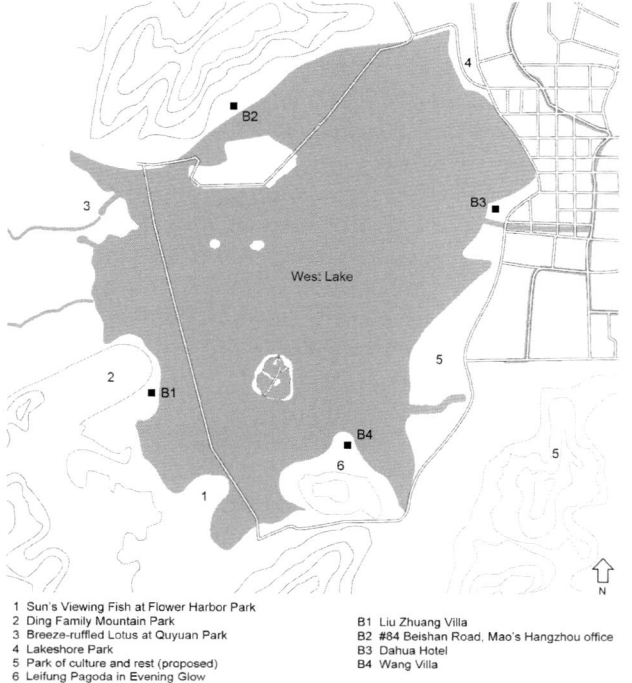

1 Sun's Viewing Fish at Flower Harbor Park
2 Ding Family Mountain Park
3 Breeze-ruffled Lotus at Quyuan Park
4 Lakeshore Park
5 Park of culture and rest (proposed)
6 Leifung Pagoda in Evening Glow

B1 Liu Zhuang Villa
B2 #84 Beishan Road, Mao's Hangzhou office
B3 Dahua Hotel
B4 Wang Villa

Figure 2.9 1950s West Lake context map, based on interview with Sun, graphic by J. Wu

Sun refers to his West Lake park design as "modern". His definition of modern was based on an earlier ideology from the Republican era that equated the notion of modern with western or foreign ideas. Sun described the design's composition as organized around five garden scenes that combined traditional Chinese gardens with western (large expanse of an open lawn based on the English garden tradition) design concepts (Figure 2.10). His proposed plantings and use of the existing topography created separation and delineated the five garden scenes within the twenty-hectare (fifty-acre) park. One of the five scenes, Garden of Harmonious Interest, *Xiequyuan*, takes its name and design inspiration from a garden scene at Beijing's Summer Palace. Sun's design also integrated the Qing dynasty monument: a stone tablet with the poetic inscriptions by the two Qing emperors. As a modernist, he made no references to the area's ecological origins. However, his design emphasized its physical and visual relationship to the lakefront. Sun's Viewing Fish at Flower Harbor park was an unusual project to have been carried out during the time, given that one of Mao's goals was to free China of foreign colonial influences and elitist ideology from traditional or imperial China.

In some ways Sun's park design reveals the deep attachment to the Chinese Picturesque tradition as an element of Chinese identity that survived even the Maoist period. Hangzhou's municipal government hired Professor Sun to design this new public park, as well as the Hangzhou Botanical Garden circa 1952–1955. Sun explained his views on the traditional Chinese gardens during the 2008 interview, "these serve as a good reference point but not something to be directly copied". Sun's intention

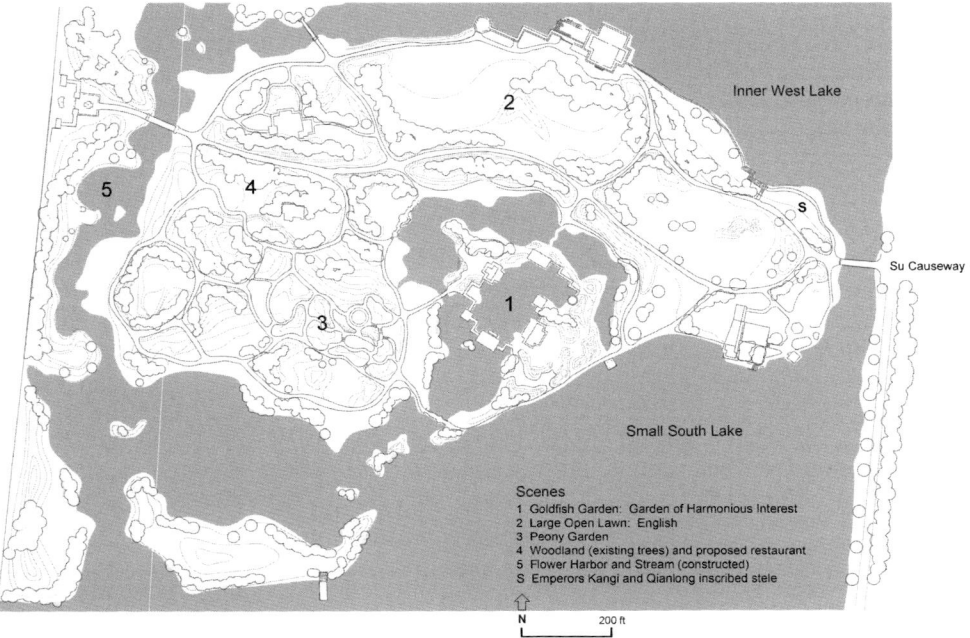

Figure 2.10 Viewing Fish at Flower Harbor Master Plan by Sun Xiaoxiang circa 1953, based on the original plan, published with permission, graphic by J. Wu

served a didactic function for CPC officials – teach them ways western garden aesthetics could be merged with Chinese design ideas. Sun described the main principle of his garden composition as "variety in unity and contrast in harmony". His use of the classical Chinese garden tradition undoubtedly helped make the park more legitimate in the eyes of Hangzhou city and CPC officials despite the shifting political context of the Mao era.

Landscape architecture education, like all educational programs and schools throughout China, mimicked the Soviet Union model during the Mao period. With Beijing as the People's Republic of China's socialist capital city, the cultural and political capital relocated from Nanjing, the ROC's capital, to Beijing. In 1951, two years after the People's Republic of China was established, faculty from the horticulture program at Peking Agricultural University and faculty from the Architectural Program at Tshinghua University established the first "gardening and greening program" at Beijing Agricultural University (Yu 2007). Beijing Agricultural University would change its name to Beijing Forestry University and the garden design program would be moved into the newly created Department of Urban and Residential Greening circa 1956–1957 (Yu 2007; Lin 2005). This coincided with Mao's Making Green the Motherland campaign and aligned with the Soviet educational model. Mao's greening campaign was seen as a precursor to the Great Leap Forward, *Da Yuejin* (GLF), program.

The launch of the GLF program in 1958 was an attempt to modernize agricultural practices in China, especially accelerate industrialization in rural areas. Mao, like many world leaders of the time, was enamored of the power of science to advance society. In Mao's case, the whole of China including its diverse natural environment, functioned as his living laboratory, particularly transforming it to serve human needs as a way to construct a socialist society. Furthermore, tensions between Mao and the Soviet leaders were growing and Mao sought to separate and demonstrate China's economic independence to the world through the GLF. Official reports at the time boldly claimed GLF's acceleration of China's industrial production beyond Great Britain within fifteen years and the USA within twenty to thirty years (Chan 2001).

Public parks throughout China were transformed into places of production during the GLF. For example, lakes, once recreational places for family boat-riding, were transformed into fish-farms; other functional recreational zones (children's play areas and the like) were converted into fruit-bearing orchards, raising pigs, or as nurseries for growing tree saplings for greening campaigns elsewhere. This conversion of public parks most likely related to Mao's "three-three system", a type of land conversion policy: one third for agriculture, specifically grain production, one third for trees or afforestation with the last third fallow. The notion of all levels of production, whether industrial, agricultural, or afforestation, was associated with the revolutionary spirit of the times (Westoby 1979; Ho 2003; Lu 2018). Mobilizing the masses through political sloganeering was typical for government officials and demonstrated loyalty to Mao.

Another activity in the GLF campaign was the "Grain First, *Yiliang weigang*" and considered the most ecologically destructive to China's natural environment (Shapiro 2000; Ho 2003). Land not suitable for grain production became a place for experimenting with local agricultural practices. Shapiro (2000) noted the extensive environmental degradation with Mao's various activities to industrialize rural China. Steel factories caused the destruction of forests and polluted river water. Mountains were transformed into productive agricultural landscapes and rivers re-aligned to produce hydro-electric power and irrigation. Few large-scale urban parks were built, and any existing parkland often doubled as a base for agricultural production. In her 1970s visit to China, Cranz (1979) was given the government's official goals for communist era parks: 1) contribute to economic productivity; 2) provide a place for workers to rest; 3) raise political consciousness; 4) popularize science; 5) show special exhibits, and 6) beautify China.

GLF's failure involved a significant decrease in grain production and contributed to the Great Famine (1959–1961) or "Three Years of Great Chinese Famine, *Sannian Da Jihuang*". Starvation from these years caused the loss of life of up to 30 million people and created major political instability (Smil 1999). Mao responded with revolutionary tactics and launched the Great Proletariat Cultural Revolution, *Wuchanjieji Wenhua Dageming*, also known as the Cultural Revolution (CR), with various sub-campaigns. A dark ten-year period ensued with the closure of schools and global isolation. Traditional cultural artifacts were violently destroyed and urban intellectuals and artists were publicly humiliated by the Red Guards during the CR's "Destroy the Four Olds and Cultivate the Four News" campaign (Spence 1991).

With the closure of schools, young people in cities were idle. The general population was also weakened by famine and significantly low economic productivity. At the same time, the perception of urban dwellers as elitist and bourgeois was revived. The combination of these social and economic factors and Mao's efforts to create unity and maintain his power base called for the launch of another campaign, "up to the mountains, down to the villages, *shanghshan xiaxiang*". Urban intellectuals and artists who were publicly humiliated by the Red Guards, and idle urban youth, *zhiqing*, were sent to the countryside for re-education and to learn from rural living.

The 1966–1976 CR period was extreme and one of the darkest times during Mao's era. The central government's official view shifted and the notion of innovation in architecture or park-building was denounced during this period. Any architecture under consideration for construction was stripped of any aesthetics. Parks were transformed into "working" landscapes for agricultural production. Cultural heritage sites and artifacts were destroyed including the loss of ancient temples, classical gardens and many ancient texts. As noted earlier, these were considered remnants of an elitist bourgeoisie denounced during the CR (Samuels 1989).

Synopsizing China's modern experience

China's experience from the 1770s through the Mao period represents a tapestry of significant change and a mosaic of modernity in four temporal stages: Early modernity, 1772–1840; Colonial modernity, 1840s–1911; Republican modernity, 1911–1949; and Mao's modernity, 1949–1978. This set the foundation for China's late 20th century hybrid modernity. The Roman Catholic Jesuit priests built European Renaissance gardens at the Old Summer Palace (destroyed in the Opium Wars) circa 1772, the Qing emperor's residence northwest of Beijing. External colonial influences introduced modernization and western modes of city-making, municipal administration and park-building for foreign consumption. Parks in China's period of colonial modernity contributed to national humiliation and suffering with the destruction of the Old Summer Palace complex of buildings and vast gardens by French and British soldiers during the second Opium War, and discrimination towards local Chinese who were excluded from public gardens and parks in foreign concession districts.

Revolution and periods of renewal, modern city-making and nation-building were intertwined with the making of public parks in the early years of the Republican era. During China's period of Republican modernity, trees and greening became symbols of liberty from China's feudal era and nearly 3000 years of imperial rule. Following the death of the founding father of modern China, Sun Yat-sen and his *Putonghua* name, Zhongshan, became emblems of national renewal, nation-building and socio-cultural re-awakening. Cities throughout China would rename their city parks and build monuments to honor Sun and celebrate a modern nation, especially during the Nanjing decade. Purple Mountain in Nanjing emerged as a modern precedent and case study for "designed landscapes", public open spaces, cultural landscapes, nation-building and modern landscape architecture. The Japanese occupation contributed to national suffering and humiliation. Domestic conflict, disunity, and guerilla warfare also caused the unraveling of the ROC and destruction of the physical environment. The rise of the peasantry and guerilla army tactics created the context for Mao Zedong's communist revolution and the establishment of the People's Republic of China in 1949 under a single party rule.

Mao's era was broadly a period of extremes and contradictions, Soviet influence, scarcity and experimentation when propaganda, sloganeering and the launching of political campaigns were part of authoritarian rule. Tree planting and greening activities from Making the Motherland Green and GLF's Three-Three system were attempts by Mao to build a socialist nation through volunteerism. Any new public parks were modeled on the Soviet park of culture. Public parks or people's parks and "red" (renamed public) squares throughout China were places for disseminating propaganda and educating the masses. Before the Great Leap Forward, the public park was a place for leisure activities where people could renew themselves to be productive citizens. West Lake, Hangzhou in China's Zhejiang Province was Mao's preferred place to retreat, conduct government business and hold meetings with CPC leaders. It would also be the place where Professor Sun's modern park was built at the site of one of West Lake's classical Ten Scenes.

The frenzied ten-month period of design and construction activities for completing the Ten Great Buildings in time for the 10th anniversary celebration of the founding of communist China set the precedent for the ad hoc design/build practices prevalent in post-Mao China and current building practices today – little time is allocated towards design while building activities begin before construction details are developed. Concurrently, the area around Tiananmen Square was bull-dozed for its expansion to accommodate some of the Ten Great Buildings and 600,000 people. As the GLF campaign was launched, public parks were converted for productivity: boating lakes became fish farms and green areas were converted for agricultural production. Natural environments in rural areas unsuited for agricultural purposes were re-purposed and ecologically degraded. Backyard steel-making polluted the natural environment. Grain production fell significantly and nationwide hunger and famine would follow. Following the three-year famine, Mao's Cultural Revolution created an aggressive authoritarian state when China's urban elite were tortured and publicly humiliated, wholesale destruction of antiquity and cultural artifacts occurred, schools closed for a ten-year hiatus, cities were depopulated as urban citizenry and the educated youth were sent to the countryside to be re-educated and learn from the peasantry. Mao's China grew economically depressed with little productivity and it became isolated from the world stage.

Narratives of modernity in China between the 1840s and 1970s involved national humiliation and suffering that overlapped westernization or colonial modernity, modernization and city-making during the Republican era and Mao's attempts to construct a socialist nation (Figure 2.11). The public park emerged initially as an imported phenomenon from the west. As China attempted to modernize during the Republican era, the public park became linked with building a modern nation and Sun Yat-sen. Trees, tree planting and greening emerged as symbols of the nation's liberty and freedom from thousands of years of feudal society, imperial rule and dynastic time. Nanjing's

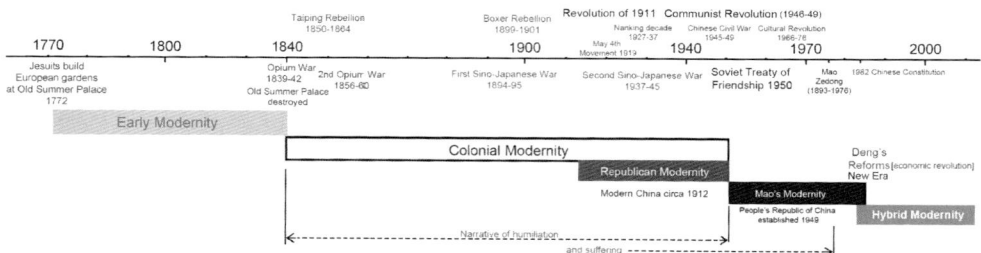

Figure 2.11 Timeline of China's four narratives of modernity traversing the 18th through the 20th centuries, graphic by author

Purple Mountain evolved as a precedent for Republican modernity and case study for modern landscape architecture with its designation as a national park (a first during China's modern time) with new spatial forms represented in the grounds of the Sun Yat-sen Mausoleum and Memorial complex, botanic gardens and national cemetery. Mao adopted the notion of greening and mobilized the masses to volunteer in tree planting activities. The Soviet park of culture and rest served as a model for new parks in the Mao era with few exceptions including Professor Sun's park in West Lake, Hangzhou. Mao's efforts to build a socialist society halted with his failed policies and political campaigns causing poverty, hunger, scarcity, the ten-year closure of schools and global isolation.

Following Mao's demise, the CPC leadership re-organized with the trial of the Gang of Four and Deng Xiaoping's rise. China's Four Modernizations and Deng's "reform and opening-up" policy triggered China's fourth 20th century (economic) revolution. Before delving into the cultural context of the late 20th century, the next chapter examines pre-modern and classical China including perspectives on nature and the environment, traditional garden culture and the design language of traditional residential private gardens – the Chinese Picturesque, a genre ingrained in contemporary Chinese culture and local identity, and a key variable in the synthesis of hybrid modernity.

References

Anderson, B. (1983) *Imagined Communities: Reflections on the Origin and Spread of Nationalism.* London, UK, Verso.

Andrews, J. & Shen, K. (2012) *The Art of Modern China.* Berkeley, CA, University of California Press.

Appadurai, A. (1990) Disjuncture and Difference in the Global Cultural Economy, *Theory, Culture, and Society*, 7(2), 295–310.

Arnold, A. (1999) *Princely Gifts & Papal Treasures: The Franciscan Mission to China & Its Influence on the Art of the West, 1250–1350.* San Francisco, CA, Desiderata Press.

Avineri, S. (1968) Marx and Modernization, *The Review of Politics*, 31(2), 172–188.

Barmé, G. (2012) A Chronology of West Lake and Hangzhou, *China Heritage Quarterly*, 28. http://www.chinaheritagequarterly.org/features.php?searchterm=028_chrono.inc&issue=028

Bays, D. (2011) *A New History of Christianity in China.* West Sussex, UK, Wiley-Blackwell.

Beardsley, J. (1998) *Earthworks and Beyond: Contemporary Art in the Landscape.* New York, NY, Abbeville Press.

Bhabba, H. K. (1994) *The Location of Culture: The Commitment to Theory.* London, UK, Routledge.

Bickers, R. A. & Wasserstrom, J. A. (1995) Shanghai's 'Dogs and Chinese Not Admitted' Sign: Legend, History, and Contemporary Symbol, *The China Quarterly*, 142, 444–466.

Bittner, S. V. (1998) Green Cities and Orderly Streets, Space and Culture in Moscow, 1928–1933, *Journal of Urban History*, 25(1), 22–56.

Bonnet, A. (1992) Art, Ideology, and Everyday Space: Subversive Tendencies from Dada to Postmodernism, *Environment and Planning D: Society and Space*, 10(1), 69–86.

Braudel, F. (1974) *Capitalism and Material Life.* New York, NY, Harper Collins.

Bray, D. (2005) *Social Space and Governance in Urban China: The Danwei System from Origins to Reform.* Stanford, CA, Stanford University Press.

Bye, A. E. Jr. (1981) *Art into Landscape: Landscape into Art.* Mesa, AZ, PDA.

Cardoso, F. H. (1979) *Dependency and Development in Latin America.* Berkeley, CA, University of California Press.

Castells, M. (1989) *The Informational City: Information Technology, Economic Restructuring, and the Urban-Regional Process*. Oxford, UK, Blackwell.

Chakrabarty, D. (2000) *Provincializing Europe: Postcolonial Thought and Historical Difference*. Princeton, NJ, Princeton University Press.

Chan, A. (2001) *Mao's Crusade: Politics and Policy Implementation in China's Great Leap Forward*. Oxford, UK and New York, NY, Oxford University Press.

Charlot, R. (1965) *The Mexican Mural Renaissance*. New Haven, CT, Yale University Press.

Cheng, T. & Selden, M. (1994) The Origins and Social Consequences of China's Hukou System, *The China Quarterly*, 139, 644–668.

Church, T. (1955) *Gardens Are for People*. New York, NY, Reinhold.

Cody, J. (2002) *Building in China: Henry K. Murphy's "Adaptive Architecture" 1914–1935*. Hong Kong, Chinese University Press.

Cody, J. (2011) Lu Yanzhi, Zhang Kaiji, and Zhang Bo - From Studio to Practice: Chinese and non-Chinese Architects Working Together. In: *Chinese Architecture and the Beaux-Arts*, Cody, J., Steinhardt, N. & Atkin, T. (eds.). Honolulu, HI, University of Hawaii Press.

Connolly, P. (1995) 101 Ideas about Big Parks, *Kerb: Journal of Landscape Architecture*, 1, 20–35

Cooper, D. (1971) *The Cubist Epoch*. New York, NY, Metropolitan Museum of Art.

Cranz, G. (1979) The Useful & the Beautiful: Urban Parks in China Landscape, *Landscape Journal*, 23, 3–10.

Cumming, E. (1997) *The Arts and Crafts Movement*. New York, NY, Thames & Hudson.

De Rachewiltz, I. (1971) *Papal Envoys to the Great Khans*. Stanford, CA, Stanford University Press.

Dirlik, A. (2002) Modernity as History: Post-Revolutionary China, Globalization and the Question of Modernity, *Social History*, 27(1), 16–39.

Dong, Y. D. (2000) Defining Beiping: Urban Reconstruction and National Identity, 1928–1936. In: *Remaking the Chinese City: Modernity and National Identity,1900–1950*, Esherick, J. W. (ed.). Honolulu, HI, University of Hawaii Press.

Doss, E. (1991) *Benton, Pollock, and the Politics of Modernism: From Regionalism to Abstract Expressionism*. Chicago, IL, University of Chicago Press.

Eckbo, G., Kiley, D. & Rose, J. (1939) Landscape Design in the Urban Environment', and 'Landscape Design in the Rural Environment. In: *Modern Landscape Architecture: A Critical Review*, Treib, M. (ed.). Cambridge, MA, MIT Press.

Eckbo, G., Kiley, D. & Rose, J. (1940) 'Landscape Design in the Primeval Environment'. In: *Modern Landscape Architecture: A Critical Review*, Treib, M. (ed.). Cambridge, MA, MIT Press.

Eckbo, G. (1950) *Landscape for Living*. New York, NY, Architectural Record with Duell, Sloan and Pearce, New York.

Edmonds, R. L. (1998) 'Studies on China's Environment'. In: *Managing the Chinese Environment*, Edmonds, R.L. (ed.). Oxford, UK, Oxford University Press.

Elvin, M. (1974) The Administration of Shanghai, 1905–1914. In: *The Chinese City Between Two Worlds*, Elvin, M. & Skinner, J. W. (eds.). Stanford, CA, Stanford University Press.

Elvin, M. (1998) The Environmental Legacy of Imperial China . In: *Managing the Chinese Environment*, Edmonds, R. L. (ed.). Oxford, UK, Oxford University Press.

Esherick, J. W. (1987) *The Origins of the Boxer Uprising*. Berkeley, CA, University of California Press.

Esherick, J. W. (2000) Modernity and Nation in the Chinese City. In: *Remaking the Chinese City: Modernity and National Identity, 1900–1950*, Esherick, J. W. (ed.), Honolulu, HI, University of Hawaii Press.

Frampton, K. (1980) *Modern Architecture: A Critical History*. London, UK, Thames & Hudson.

Frank, A. G. (1969) *Capitalism and Underdevelopment in Latin America*. London, UK, Penguin.

Gehl, J. (1971, trans. 1987) *Life* Between *Buildings: Using Public Space.* Copenhagen, Denmark, The Danish Architectural Press.

Giddens, A. (1990) *The Consequences of Modernity.* Stanford, CA, Stanford University Press.

Giedion, S. (1977) *Space, Time and Architecture: The Growth of a New Tradition.* Cambridge, MA, Harvard University.

Gough, M. (2005) *The Artist as Producer: Russian Constructivism in Revolution.* Berkeley, CA, University of California Press.

Greenburg, C. (1989) *Art and Culture: Critical Essays.* Boston, MA, Beacon Press.

Halprin, L. (1969) *The RSVP Cycles: Creative Processes in the Human Environment.* New York, NY, George Braziller.

Halprin, L. & Burns, J. (1974) *Taking Part: A Workshop Approach to Collective Creativity.* Cambridge, MA, MIT Press.

Harden, J. D. (1995) Liberty Caps and Liberty Trees, *Past and Present*, 146(1), 66–102.

Harrison, C. & Wood, P. eds. (1992) *Art in Theory: 1900-2000.* Oxford, UK, Blackwell Publishing.

Hayden, P. (2005) *Russian Parks and Gardens.* London, UK, Frances Lincoln.

Ho, P. (2003) Mao's War Against Nature? The Environmental Impact of the Grain-First Campaign in China, *The China Journal*, 50, 37–59.

Hsai, P. F. (1997) 'Chinoiserie' in Eighteenth Century England, *American Journal of Chinese Studies*, 4(2), 238–251.

Hung, C. (2013) A Political Park: The Working People's Cultural Palace in Beijing, *Journal of Contemporary History*, 48(3), 556–577.

Hung, W. (1991) Tiananmen Square: A Political History of Monuments, *Representations*, 35(1), 84–117.

Inkeles, A. & Smith, D. H. (1974) *Becoming Modern: Individual Change in Six Developing Countries.* Cambridge, MA, Harvard University Press.

Jackson, J. B. (1970) *Selected Writings by J. B. Jackson.* Amherst, MA, University of Massachusetts Press.

Jackson, I. (2017) Habitability in the Treaty Ports: Shanghai and Tianjin. In: *The Habitable City in China: Urban History in the Twentieth Century*, Lincoln, T. & Xu, T. (eds.). London, UK, Palgrave Macmillan.

Jacobs, J. (1961) *The Death and Life of American Cities.* New York, NY, Random House.

Jacques, D. (1990) On the Supposed Chineseness of the English Landscape Garden, *Garden History*, 18(2), 180–191.

Jencks, C. (1973) *Modern Movements in Architecture.* London, UK, Penguin.

Jencks, C. (2002) *New Paradigm in Architecture.* New Haven, CT, Yale University Press.

Johnston, C. (ed.) (2006) *Seeing High and Low: Representing Social Conflict in American Visual Culture.* Berkeley, CA, University of California Press.

Kaprow, A. (1966) *Assemblage, Environments & Happenings.* New York, NY, Harry N. Abrams.

Kellein, T. (2002) *Donald Judd: Early Works 1955–1968.* New York, NY, D.A.P.

Kerr, C., Dunlop, J., Harbison, F. & Myers, C. (1964) *Industrialism and Industrial Man.* New York, NY, Oxford University Press

Keswick, M. (1978) *The Chinese Garden: History, Art and Architecture.* London, UK, Academy.

Kiley, D. & Amidon, J. (1999) *Dan Kiley: The Complete Works of America's Master Landscape Architect.* Boston, MA, Little, Brown and Company.

Lai, D. (2005) Searching for a Modern Chinese Monument: The Design of the Sun Yat-sen Mausoleum in Nanjing, *Journal of the Society of Architectural Historians*, 64(1), 22–55.

Lee, E. (1930) The Three Principles of the Kuomintang, *Annals of the American Academy of Political and Social Science*, 152(1), 262–265.

Levine, N. (1996) *The Architecture of Frank Lloyd Wright*. Princeton, NJ, Princeton University Press.

Levenson, J. (1968) *Confucian China and Its Modern Fate: A Trilogy*. Berkeley, CA, University of California Press.

Levy, M. J. (1966) *Modernization and the Structure of Societies: A Setting for International Affairs*. Princeton, NJ, Princeton University Press.

Lewis, J. W. (1963) *Leadership in Communist China*. Ithaca, NY, Cornell University Press.

Lin, G. S. (2005) Review and Prospect: A Study of Landscape Architecture Education in China, *China Landscape Architecture Journal*, 9 and 10, 1–9, 73–79, translated.

Lu, H. (2018) Shanghai Flora: The Politics of Urban Greening in Maoist China, *Urban History*, 45(4), 660–681.

Lüthi, L. (2008) *The Sino-Soviet Split: Cold War in the Communist World*. Princeton, NJ, Princeton University Press.

Marcus, C. C. (1974) *Children's Play Behavior in a Low Rise, Inner City Housing Development*. Berkeley, CA, Institute of Urban and Regional Development, University of California at Berkeley.

McHarg, I. (1969) *Design with Nature*. New York, NY, Natural History Press.

Mertin, D. (2014) *Mies*. London, UK and New York, NY, Phaidon.

Musgrove, C. D. (2013) *China's Contested Capital: Architecture, Ritual, and Response in Nanjing*, Honolulu, HI, University of Hawaii Press.

Padua, M. G. (2014) China: New Cultures and Changing Urban Cultures. In: *New Cultural Landscapes*, Roe, M. & Taylor, K. (eds.). Abingdon, UK and New York, NY, Routledge.

Pevsner, N. (1936) *Pioneers of the Modern Movement*. London, UK, Faber & Faber.

Profous, G. (1992) Trees and Urban Forestry in Beijing, China, *Journal of Arboriculture*, 18(3), 145–154.

Reid, S. (2001) Socialist Realism in the Stalinist Terror: The Industry of Socialism Art Exhibition, 1935–41, *The Russian Review*, 60(2), 153–184.

Richardson, S. D. (1990) *Forests and Forestry in China*. Washington, DC, Island Press.

Rose, J. C. (1958). *Creative Gardens*. New York, NY, Reinhold Publishing Corporation.

Rowe, P. & Seng, K. (2002) *Architectural Encounters with Essence and Form in Modern China*. Cambridge, MA, MIT Press.

Ruan, X. (2011) Yang Tingbao, China's Modern Architect in the Twentieth Century. In: *Chinese Architecture and the Beaux-Arts*, Cody, J., Steinhardt, N. & Atkin, T. (eds.). Honolulu, HI, University of Hawaii Press.

Said, E. (1977) *Orientalism*. New York, NY, Pantheon Books.

Samuels, M. (1989) Beijing and the Power of Place in Modern China. In: *The Power of Place: Bringing Together Geographical and Sociological Imaginations*, Agnew, J. & Duncan, J. (eds.). Winchester, MA, Unwin Hyman Ltd.

Sandler, I. (2009) *Abstract Expressionism and the American Experience: A Reevaluation*. Manchester, VT, Hudson Hills.

Schlesinger, A. (1952) Liberty Tree: A Genealogy, *The New England Quarterly*, 25(4), 435–458.

Shapiro, J. (2000) *Mao's War Against Nature: Politics and the Environment in Revolutionary China*. Cambridge, UK, Cambridge University Press.

Shi, M. (1998) From Imperial Gardens to Public Gardens: The Transformation of Urban Space in Early 20th Century Beijing, *Modern China*, 24(3), 219–254.

Skinner, G. W. (ed.) (1977) *The City in Late Imperial China*. Stanford, CA, Stanford University Press.

Smil, V. (1984) *The Bad Earth: Environmental Degradation in China*. London, UK, Zed Books.

Smil, V. (1999) China's Great Famine: 40 Years Later, *British Medical Journal*, 319(7225), 1619–1621.

Sontag, S. (1966) *Against Interpretation*. New York, NY, Farrar, Straus, Giroux.

Spence, J. D. (1991) *The Search for Modern China*. New York, NY, W. W. Norton & Company.

Spence, J. D. (1996) *The Chinese Son: The Taiping Heavenly Kingdom of Hong Xiuquan*. New York, NY, W.W. Norton.

Sun, X. (1994) The City Should Be Rich in the Pleasures of Wild Nature – A Traditional Aesthetic Concept of China for Urban Planning, *Ekistics*, 61(364/365), 22–28.

Taylor, J. (1968) Futurism: Dynamism as the Expression of the Modern World Introduction. In: *Theories of Modern Art*, Chipp, H. (ed.). Berkeley, CA, University of California Press.

Thomas, G. (2009) Yuanming Yuan/Versailles: Intercultural Interactions between Chinese and European Palace Cultures, *Art History*, 32(1), 115–143

Treib, M. (1993) Axioms for a Modern Landscape Architecture. In: *Modern Landscape Architecture: A Critical Review*, Treib, M. (ed.). Cambridge, MA, MIT Press.

Tunnard, C. (1938) *Gardens in the Modern Landscape*. London, UK, The Architectural Press.

Wagner, R. (2011) Ritual, Architecture, Politics and Publicity during the Republic. In: *Chinese Architecture and the Beaux-Arts*, Cody, J.., Steinhardt, N. & Atkin, T. (eds.). Honolulu, HI, University of Hawaii Press.

Waldheim, C. ed. (2006) *The Landscape Urbanism Reader*. New York, NY, Princeton Architectural Press.

Walker, P. & Simo, M. (1994) *Invisible Gardens: The Search for Modernism in the American Landscape*. Cambridge, MA, MIT Press.

Wallerstein, I. (1974) The Rise and Future Demise of the World Capitalist System, *Contemporary Studies in Society and History*, 16(4), 387–415.

Wang, H. (2003) *China's New Order: Society, Politics, and Economy in Transition*. Cambridge, MA, Harvard University Press.

Wasserstrom, J. N. (2000) Locating Old Shanghai: Having Fits about Where It Fits. In: Esherick, J. W. (ed.). *Remaking the Chinese City: Modernity and National Identity, 1900–1950*, Honolulu, HI, University of Hawaii Press.

Westoby, J. C. (1979) "Making Green the Motherland" Forestry in China. In: *China's Road to Development*, Maxwell, N. (ed.). Oxford, UK, Pergamon.

Whyte, W. (1980) *The Social Life of Small Urban Places*. Washington DC, The Conservation Foundation.

Wong, Y. T. (2001) *A Paradise Lost: The Imperial Garden Yuanming Yuan*, Honolulu, HI, University of Hawaii Press.

Wright, A. F. (1977) The Cosmology of the Chinese City. In: *The City in Late Imperial China*, Skinner, G. W. (ed.). Stanford, CA, Stanford University Press.

Xing, R. (2002) Accidental Affinities: American Beaux-Arts in Twentieth-Century Chinese Architectural Education and Practice, *Journal of the Society of Architectural Historians*, 61(1), 30–47.

Xu, Y. (2000) *The Chinese City in Space and Time*, Honolulu, HI, University of Hawaii Press.

Xue, L. (2008) The Grand Bandstand, *Huazhong Architecture*, 26, 179–190, translated.

Yu, K. (2007) The Evolution of Landscape Architecture in China. In: *Proceedings of Between Architecture and Landscape Education*, Laffage, A. (ed.). Paris, France, Ecole Nationale Superieure d'Architecture de Paris la Villette, 22–23 Nov. 2007.

Yu, K., Li, D. & Li, N. (2006) The Evolution of Greenways in China, *Landscape and Urban Planning*, 76(1–4), 223–239.

Zhao, J. & Woudstra, J. (2012) 'Making Green the Motherland': Greening the Chinese Socialist Undertaking (1949–1978), *Studies in the History of Gardens and Designed Landscapes*, 32(4), 312–330.

Zou, H. (2011) *A Jesuit Garden in Beijing and Early Modern Chinese Culture*. West Lafayette, IN, Purdue University Press.

3 Pre-modern China

Nature, cosmology, mythology and the Chinese Picturesque

"The wise find pleasure in water; the virtuous find pleasure in hills".

Confucius Analects

The "garden" is the companion archetype to the public park in the discipline of landscape architecture. As a world civilization, China developed its own distinctive garden centuries ago. This chapter introduces China's traditional garden and related culture along with pre-modern concepts on nature, cosmology, folklore and mythology, and religious influences. China's pre-modern period spans ancient times through the classical imperial era before British influence circa 1840s and colonial modernity. China's pre-modern cultural and religious phenomena served as inspirations for the garden's meaning and spatial forms, especially in the historical development of China's private residential garden – Scholar Garden, also known as Gardens of the Literati, *wenren yuanlin*. Essentially, these were amateur gardens designed by and for retired scholar-officials from China's imperial court. The accepted typology of China's traditional designed landscapes includes the larger royal gardens, temple gardens and private residential gardens.

Understanding China's traditional park and garden culture and their spatial forms is important to this research for two reasons. First, it is necessary in order to understand the degree to which the four late 20th century case study parks represent a departure from previous work or represent a continuation of established traditions. Second, these traditional gardens have served as a source of some of the key elements of a design language that has been linked to Chinese identity. The author refers to this traditional design genre as the Chinese Picturesque, appropriating from Hunt's (2002) discursive narrative on the English Picturesque garden. More critically, these traditional gardens serve as a source for "local" culture and identity, a key variable in the synthesis of post-Mao China's hybrid modernization and related hybrid modernity.

Research for this chapter draws from the literature, interviews with Chinese designers, educators and scholars. It also involved field research in China including numerous visits to traditional Chinese gardens in Suzhou, Jiangsu Province and local region, West Lake in Hangzhou, Zhejiang Province, as well as to the larger former royal parks in and around Beijing including the summer resort in Chengde, Hebei Province. It is widely accepted that Suzhou was the garden capital in the Ming and Qing dynasties where numerous Chinese private residential gardens were built by retired government scholar-officials. The scholarship on these gardens is significant and the goal for this chapter is to introduce key concepts for understanding the Chinese Picturesque genre and related spatial forms. For further studies, the author recommends: Chen's (1984) *On Chinese Gardens*; Keswick's (1978) *The Chinese Garden*; Hardie's (1988) English translation of Ji Cheng's (1631) *The Craft of Gardens*; and a series of essays on Chinese gardens in two special issues of the journal *Studies in the History of Gardens & Designed Landscapes* (1998, 1999). *Art in China* provides an excellent companion to understanding China's history of visual arts and culture (Clunas 1997). Li's translated work (2010), *The Chinese Aesthetic Tradition*, provides a historical overview and general foundation for understanding the cultural evolution of aesthetics in China. Fairbank's edited book (1957) entitled *Chinese Thought and Institutions* offers a series of rich interpretations on the political struggles of Confucian ideology and classical texts in China's socio-political institutions including a socio-political history of China for the novice. Wright's (1977) book chapter, "Cosmology of the City", sets the benchmark for understanding the morphology and evolution of ancient and imperial Chinese cities. It is widely accepted that architecture in China evolved as a result of the Chinese traditional garden.

Most critical to understanding the difference between China and western culture is the notion of ancestral worship and ways this played out in ancient sacred landscapes. For example, earth-mounds initially surrounded by a moat of water were a place of ancient ancestral worship. Over time, these sacred places were replaced by ancestral halls, *ci miao* or *ci tang*, buildings containing stones or stelae inscribed with calligraphy and poetry by and for ancestors. Another contrast is the western Judeo-Christian notion of "suffering" on earth with paradise found in heaven in the afterlife. In Chinese folklore and mythology, and later religious traditions, their beliefs were "of this world", where paradise can be found on earth in nature or "miniaturized nature" in traditional gardens.

The focus of this chapter is the inherited form and symbolism of the Chinese traditional garden, a major actor for the Chinese Picturesque. It is vital to understand the Chinese Picturesque genre and the way this garden tradition historically developed. It also is important to "locate" it within the larger Chinese cultural context. The meaning of nature in China is related to the evolution of the generally accepted typology of designed landscapes mentioned earlier: the larger imperial parks, temple gardens and smaller residential private gardens. The evolution of traditional China's park and garden culture is directly linked to cosmology, rituals, mythology and folklore, and essentially parallels cultural development for this world civilization. In the early imperial era, religion (Buddhism, Daoism) and Confucian philosophy were intertwined with the arts and life among China's elite, as well as feudal society.

The Gardens of the Literati have a distinctive place in China's cultural development and are considered an extension of the four scholar arts, *siyu*, practiced by scholars in the classical period: *qin*, playing the stringed musical instrument known as *guqin*; *qi*, playing a board game; *shu*, calligraphy; and *hua*, Chinese painting. These four arts can be understood as representing skills in music, strategy or the discipline of the mind, visual poetry, and the fine art of painting. Like landscape painting and calligraphy, the traditional scholar-official garden continues to be seen as one of the vessels of Confucian Chinese identity. References to the garden design tradition remain a means of marking the Chinese character of a designed landscape, and for some contemporary practitioners, highly contested.

Nature, cosmology, mythology and folklore

Unlike other civilizations, China's pre-modern era, particularly ancient times, had a primordial view of nature or the notion of "wilderness". Divine spirits in ancient China corresponded with real world natural phenomena, e.g. trees, mountains, stones and animals, in addition to notions of heaven and earth. The Classic of Mountains and Seas, *Shanhaijing*, is widely accepted ancient writing that serves as a major source of Chinese mythology. It provides a rich narrative of China's earliest ancestors and their rites, rituals and mythical figures of the period spanning the 3rd century BCE to the 2nd century of the CE. Dragon worship, the mythical Kunlun Mountains and the "lands beyond the seas" are covered in this book, as are ancient cosmography, interpretations of ancient geography, folk medicine, mythical animals, deities and China's earliest ethnic tribes (Birrell 1999). It discusses the perception that the world or China was in the form of a two-dimensional square and part of the celestial universe. In this narrative, China, the Middle Kingdom, a sacred place and geometrically square, was surrounded and separated on four sides by non-Chinese peoples and foreign lands. Heaven or the sky overhead was round and held up by massive mountains functioning as pillars rising out of the square-shaped earth. The Five Deities, *Wudi*, evolved out of this mythology and became part of nature worship in ancient China; the five mountain gods were geographically located in the east (*Taishan*, Tranquil Mountain, Shandong Province), south (*Hengshan*, Balancing Mountain, Hunan Province), west (*Huashan*, Splendid Mountain, Shaanxi Province), north (*Hengshan*, Permanent Mountain, Shanxi Province) and center (*Songshan*, Lofty Mountain, Henan Province). Over time, these Five Great Mountains, *Wuyue*, were physically realized with temples built at each location where emperors worshipped.

Given the emperor was considered the "son of heaven", it was a major responsibility to visit the sacred mountains regularly to appease the spirits, gods and ancestors. Otherwise, the prosperity and well-being of the entire empire were at risk. Grand tree-lined roads, or the "imperial way, *yudao*", were constructed for the emperor and his entourage to approach the sacred mountains. Steps or "staircases to heaven" were carved into the mountains' slopes for the emperors to ascend the summit. Over time, these natural sacred mountain environments developed into popular pilgrimage sites and scenic areas for national consumption. The natural landscape in the *Classics of Mountains and Seas* is expressed in terms of its abundance and contributes to China's ancient folklore and mythology as its own paradise. The natural landscape is also described with deep reverence including details of the sensory aspects and medicinal quality of rare plants, as well as visualizations of gold, silver and jade. Furthermore, narratives on mountains, deities and mythical animals living in fairyland from *The Classics of Mountains and Seas* were inspirations for China's earlier park and garden-making traditions.

It is widely accepted among Chinese garden historians that the first known recreation and hunting preserve was a large enclosed tract of land designated by Western Zhou's first ruler. Referred to as *you*, this animal preserve with agricultural lands contained an altar, *tai*, on a built-up earth mound, and a pond, *zhao* (Sun 1984; Feng 1992). The Book of Changes, *I-Ching*, the oldest Chinese classic text, noted human sacrifices to deities were part of rituals that would take place at this altar, usually before military battles or in advance of agricultural harvests during ancient times (Feng 1992).

As ancient China evolved, the philosophical concept of "harmony with nature, *rong he*", or "unity of man with nature", was an accepted societal norm. The Chinese word for nature, *da jiran*, literally translates into "everything coming into being" and expresses the totality of mountains, rivers, plants, animals and humans and is reflective of the transformative nature of the five elements, *wu xing*: wood, *mu*; fire, *huo*; earth, *tu*; metal, *jin*; and water, *shui*. Both the religion, Daoism, and Confucian philosophy were sources of China's pre-modern notion of man's harmonious co-existence with nature. The writings of Lao Zi, author of the sacred text Book of the Way, *Dao De Jing*, written circa 770–476 BCE, emphasized the importance of "nature" as the ruler of the world, not man (Sun 1984). Lao's writings on Daoism noted that human beings should imitate the earth, while earth imitates the universe and the universe imitates *Dao*, the "way". In the essence of Daoism, human beings co-exist with nature and man should harmonize with nature's cycles and rhythms (Tang 2015).

The view on "harmony with nature" by Chinese philosopher Confucius, also known as *Kong Qiu*, draws from the Book of the Way in his classical books, The Doctrine of the Mean and The Analects of Confucius (Sun 1984). Confucius noted the importance of "balance, *ping heng*" as a basic principle and "harmony, *he*" as the primary order of the world (Sun 1994). Confucius also believed in the nurturing dimensions of nature – "to seek ultimate wisdom in nature", and nature's relationship to society's health and well-being (Chen & Wu 2009). Later, during China's imperial era, these Confucian texts and other classical works were vital knowledge needed to pass the civil service imperial examinations. In ancient and imperial times, China was less interested in dominating nature in their garden and park traditions as did their western and Christian counterparts. With Daoism and especially the Confucian-dominated imperial system, the concepts of harmony with nature and learning from nature were entrenched in pre-modern China. Over time, Chinese society adopted this notion as an anthropocosmic perspective of the world (Weller 2011).

Like other civilizations, pre-modern China had its own utopian landscape and visions of paradise. These were also subjects and inspirations for poetry and paintings, as well as for garden designers. One type of paradise, known as "Peach Blossom Spring", was first introduced around the Jin dynasty, a period of civil unrest and disunity, by the scholar and poet, Tao Qian, also known as Tao Yuanming, in an essay with the same title (Tao 1965). It is comparable to the ancient Greek narrative for Arcadia. The Chinese folkloric version describes a fisherman in an idyllic unspoiled natural environment who observes people living harmoniously in a picturesque landscape comprised of mountains and fertile valleys; and his journey includes a sensational immersive experience along a stream with peach trees blossoming on both sides of the riverbank. The revelatory experience of moving through this visually provocative environment with people living contently within their surroundings, along with the culminating sensation of being overcome by the beauty of peach trees blossoming in the spring, emerged as an idyllic landscape image.

Another vision of paradise, "world in the bottle gourd", originated with another philosopher, Ge Hong (283–343 CE), who is known to have reconciled Daoism and Confucianism and lived during China's period of disunity. This paradisiac narrative is fantasy-based with fairyland as a place where "immortals" or gods live. It describes the daily life of an immortal who lived in a bottle-shaped gourd; he exited the gourd to sell elixirs daily and returned to the gourd to sleep each night. In this narrative, a curious person follows the immortal into the gourd to discover a grand picturesque landscape of unearthly beauty (Yu & Tredici 1993). Garden designers would later draw parallels of "miniaturized landscapes" and experiencing a "small" enclosed space as a transformative method to expand the feeling of space.

The "world in a pot" prototype is considered another folkloric reference for the smaller, enclosed, private scholar-official garden versus the larger imperial or palace building and garden complexes built for Chinese emperors. This particular prototype is inspired by China's birthplace, the Yellow River basin, also known as the Guanzhong region. Spatially, the shape of the area represents a giant pot with the surrounding mountains acting as the walls of the pot, with the adjacent Wei River basin as the pot's interior (Chen & Wu 2009). In this narrative, the Yellow River and its tributaries serve as the corridor through which exchanges with the rest of the world took place. In effect, scholars have interpreted this geographic metaphor as an explanation for a culture that favors enclosed spaces.

Feng-shui (wind-water) theory, based on cultural and religious beliefs, was an important cosmological practice in pre-modern China. In basic terms, it is a set of "harmonious" site selection principles for locating and designing dwellings and burial places. In more complex cultural terms, it draws from various abstract concepts: Daoist dualism, known as *Yin-Yang*, negative-positive (female-male); Five Elements (metal, wood, water, fire, earth) *Wu Xing*; and Eight Trigrams, *Bagua*. Each of these concepts deals with transformation, inter-relationships and types of ordering systems to achieve harmony and eternity of the whole (Chen & Wu 2009). Essentially, *feng-shui* practice itself deals with creating harmony between an occupant and the built dwelling, including sacred burial grounds, through the manipulation of "*Qi*", the life force or energy that drives all change.

Two "schools" of thought have been widely accepted among scholars as *feng-shui* practices: the "form" school (Figure 3.1) manipulates *qi* through understanding the relationship between landscape features (mountains are imagined as the dragon form and flowing water symbolizes prosperity) and the movement of *qi*; and the "compass" school which analyzes *qi* through the use of a geomancer's instrument combined with an analysis of astrological changes (Tan 2011). *Feng-shui* practices were somewhat indirect in the design of scholar gardens but fundamental to the basic siting of cities, villages and dwellings, and burial sites (Johnston 1991). Since emperors in China were considered the Son of Heaven, *Tianzi*, burial sites for emperors were considered sacred.

When China was unified in the Qin dynasty, its first emperor, Qin Shi Huang, organized the construction of Shang-ling Park. Comprised of several palace buildings, the spatial form of this designed landscape incorporated a lake with three artificially made islands. Apparently, the lake with three islands represented a type of fairyland where the immortals lived (Sun 1984; Keswick 1978). The spatial model of the lake with three islands would continue and its meaning would change with the times (Feng 1992).

Brief history of China's designed landscapes

The history of designed landscapes in China stretches back thousands of years. Three general types of enclosed outdoor spaces and designed landscapes were created in China during the period that spans 255 BCE to 1911 CE (the Qin through the Qing dynasties), respectively: the large imperial park, *huang jia yuanlin*; the temple garden, *si yuan yuanlin*; and the residential garden, *si jia yuanlin*. All three types are still evident in China today. A fourth type – the imperial hunting grounds, *you*, or *ling you*, (unenclosed hunting preserve outside the walled imperial city) mentioned earlier – has ceased to

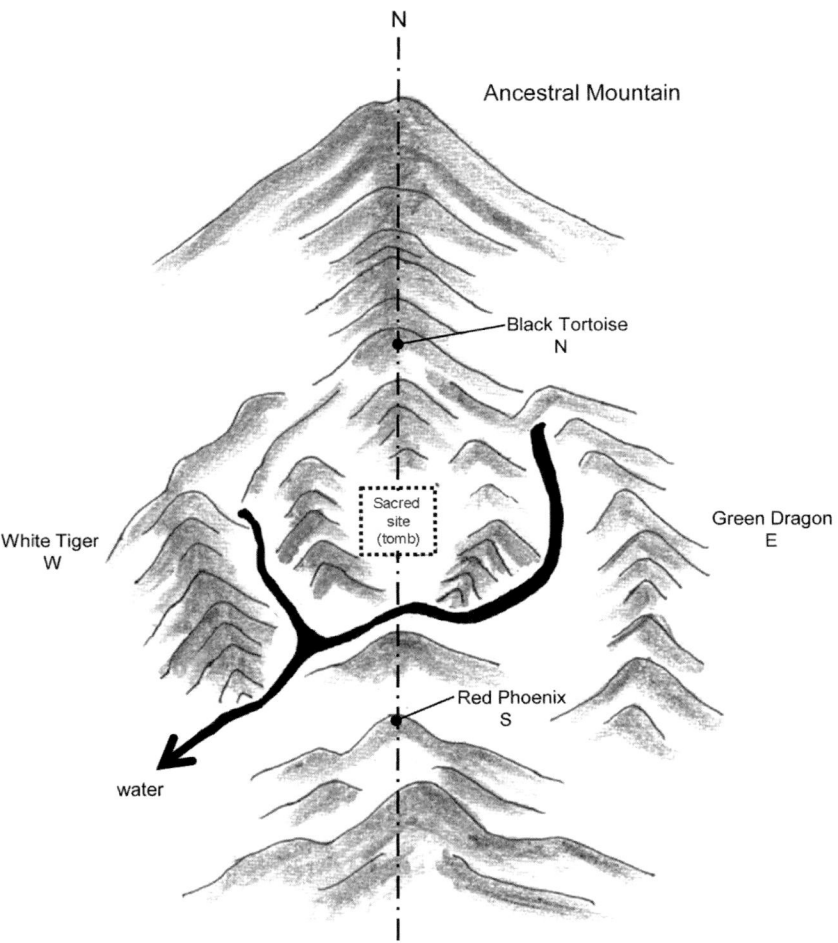

N
|
Ancestral Mountain

Black Tortoise
N

Sacred
site
(tomb)

White Tiger
W

Green Dragon
E

Red Phoenix
S

water

Figure 3.1 Feng-shui "form" school cosmological site planning, graphic by author

exist. However, scholars have found references to this fourth type in ancient and classical Chinese literature, as well as in paintings (Clunas 1997; Keswick 1978; Feng 1992).

Two types of designed landscapes were typically found at the larger properties for Chinese emperors: the outdoor spaces in and around the emperor's palace, and scenic parks, *feng jing qu*. Examples of imperial parks in and around the emperor's residences include the garden spaces within Beijing's imperial palace complex of the Forbidden City, *Zijincheng*, and surrounding parks, Northern Sea, *Beihai*, and Prospect Hill, *Jingshan*. Within proximity of Beijing's imperial palace and grounds were two other royal parks: the old summer palace, the Garden of Perfect Brightness, *Yuanming Yuan*, currently in ruins, and the Garden of Nurtured Harmony, *Yihe Yuan*, both northwest of the Forbidden City. Fleeing-the-Heat Mountain Villa, *Bishu Shanzhuang*, another large-scale imperial pleasure ground covering 5.5 square kilometers (around 2.1 square miles) was developed by Qing dynasty emperors in Chengde, Hebei Province. This summer resort property is located 250 kilometers northeast of Beijing and the Qing dynasty

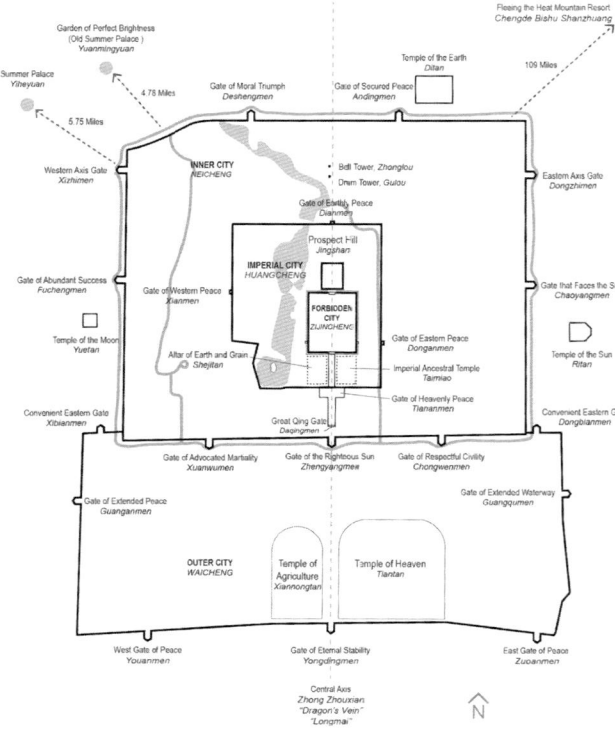

Garden of Perfect Brightness
(Old Summer Palace)
Yuanmingyuan

Fleeing the Heat Mountain Resort
Chengde Bishu Shanzhuang

Summer Palace
Yiheyuan

109 Miles

4.78 Miles

Temple of the Earth
Ditan

Gate of Moral Triumph
Deshengmen

Gate of Secured Peace
Andingmen

5.75 Miles

INNER CITY
NEICHENG

Bell Tower, *Zhonglou*
Drum Tower, *Gulou*

Eastern Axis Gate
Dongzhimen

Western Axis Gate
Xizhimen

Gate of Earthly Peace
Dianmen

Prospect Hill
Jingshan

IMPERIAL CITY
HUANGCHENG

Gate of Abundant Success
Fuchengmen

Gate of Western Peace
Xianmen

Gate that Faces the Su
Chaoyangmen

FORBIDDEN
CITY
ZIJINCHENG

Temple of the Moon
Yuetan

Altar of Earth and Grain
Shejitan

Gate of Eastern Peace
Donganmen

Imperial Ancestral Temple
Taimiao

Temple of the Sun
Ritan

Convenient Eastern Gate
Xibianmen

Great Qing Gate
Daqingmen

Gate of Heavenly Peace
Tiananmen

Convenient Eastern G
Dongbianmen

Gate of Advocated Martiality
Xuanwumen

Gate of the Righteous Sun
Zhengyangmen

Gate of Respectful Civility
Chongwenmen

Gate of Extended Peace
Guanganmen

Gate of Extended Waterway
Guangqumen

OUTER CITY
WAICHENG

Temple of
Agriculture
Xiannongtan

Temple of Heaven
Tiantan

West Gate of Peace
Youanmen

Gate of Eternal Stability
Yongdingmen

East Gate of Peace
Zuoanmen

Central Axis
Zhong Zhouxian
"Dragon's Vein"
"*Longmai*"

N

Figure 3.2 Location map of Imperial Gardens in and around Beijing, graphic by X. Liu

emperors had it built to escape the summer heat (Figure 3.2). They also conducted government business there during the summer months (Keswick 1978; Johnston 1991).

Developed as a summer resort, this Fleeing-the-Heat Mountain Villa or Resort included large- scale recreational areas, artificial lakes and islands, artificial and natural mountains, several scenic areas and replica architecture. When originally constructed, these large royal parks were typically enclosed by defensible walls (Keswick 1978; Graham 1938; Siren 1949; Valder 2002). Mimetic practices or the copying of gardens and buildings were typical throughout China. For example, a smaller version of the Tibetan Potala Buddhist Temple complex was built at the Chengde summer resort. Versions of West Lake were built throughout imperial China. These mimetic design practices also took place after Deng's late 20th century reforms. Examples include Le Corbusier's Ronchamp Chapel in Zhengzhou, Henan Province built circa 1994; Orange County, Beijing suburbs – a replica of a southern California single-family residential development built circa 2002; Eiffel Tower in "Little Paris" near Shanghai built in 2007; and London's Tower Bridge in Suzhou, Jiangsu Province completed in 2012 (Yu & Padua 2007; Bosker 2013).

Another type of large royal park was the scenic park, *feng jing qu*, vast natural areas chosen for their visual and cultural value. For example, the Yellow Mountain Scenic Area, *Huangshan Fengjing Mingshenqu* in Anhui Province, essentially a natural conservation area protected by the Song dynasty Emperor Quinzong circa 1125 CE, and a type of scenic park (Keswick 1978; Valder 2002). Yellow Mountain and its surrounding natural scenery were a source of inspiration for classical artists and poets and the subject of many works of art. The Yellow Mountain area is believed to be the first known natural and cultural area designated by an emperor for conservation purposes (Steinitz 2008). The closest equivalent park type outside China are the natural or wilderness parks in the USA or UK national park systems. Yellow Mountain Scenic Area and several of the former royal parks and gardens have been designated as major sites of historic, cultural and natural significance in China and registered on UNESCO's list of World Heritage sites.

Temple or religious gardens, *si yuan yuanlin*, existed in traditional China among the various religions and sects of Buddhism, Taoism and other local religions. These generally followed the traditional courtyard architectural lay-out called for in a pre-Qin classical text for city-making theory called the Rites of Zhou, *Zhouli* (Wright 1977). These outdoor landscapes were part of religious temple complexes rather than formal pleasure or recreational grounds. The outdoor temple landscapes were also functional spaces where monks lived and conducted religious activities (Keswick 1978; Valder 2002).

As Chinese society culturally evolved through the long imperial era, temple (Confucian, Taoist, Buddhist) grounds functioned as community public space and social centers, as well as religious centers. In addition to acting as circulation space and destinations for the daily religious and residential activities of monks, these grounds served as locations for temple fairs. These periodic fairs were typically held during religious celebrations in both urban and rural areas with popular entertainment and commercial activities including magicians, acrobats, and commercial market stalls (Zhao 2002; Wang 1999). Temple fairs in rural villages usually took place when the community celebrated local deities according to the agricultural timetable, particularly during harvest and planting seasons (Zhao 2002).

The third type of designed landscape was the private residential garden and involved two spatial forms: the residential courtyard, *sen lin*, and the private garden, *sijia yuanlín*. The residential courtyard is the common outdoor space within a walled family compound with its residential courtyard dwelling unit known as *siheyuan*, "four-sided enclosed courtyard"; and the courtyard space is known in Chinese as *tingyuan*. The dwelling complex typically had a north-south orientation and was formed by inward-facing buildings on four sides around a common courtyard (Knapp 1989). The north-south orientation is based on the cosmology of city-making laid out in the Rites of Zhou, *Zhouli*. Private residential gardens were designed by and for scholar-officials, *wenren*, or literati, who were the elite class and members of the Chinese imperial government. They developed their personal private gardens (Scholar Gardens or Gardens of the Literati) into an art form after they retired from government service.

Interpreting garden-making throughout pre-modern China

The meaning of imperial parks and scholar gardens changed with the times and are emblematic of both China's cultural development and the Chinese Picturesque. Feng (1992) succinctly describes five temporal and overlapping periods; and each period was elucidated with one word, respectively: "Form, Emotion, Reasoning, Spirit and Intention". An overview in Figure 3.3 elucidates Feng's interpretation of royal parks and gardens for each period. It includes his speculations on the evolution of thinking: object to subject; coarseness to the more refined; and from superficial to deep meaning; and his essay draws from Chinese literature, poetry and the visual arts of the related period.

Feng's (1992) first period, Form, discusses the basic morphology of the large royal parks built during the ancient Warring States period: a lake and three artificially made

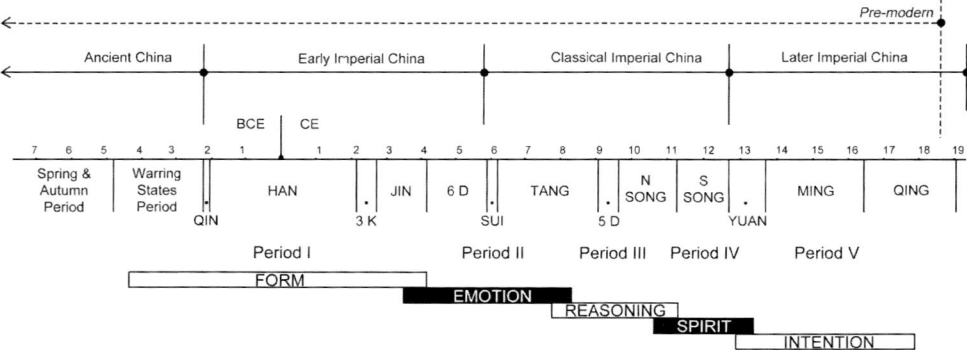

Figure 3.3 Historic garden periods based on Feng (1992), graphic by J. Wu

islands where immortals lived. In this context, nature is copied and miniaturized, and symbolizes paradise. Feng's (1992) second period, "Emotion, spans part of China's Early Imperial period through the Tang dynasty. The influence of Buddhism was significant during this period and many successful and rich merchants donated their lavish properties (mansions with gardens) for use as temples. At the same time, it was a period of civil unrest with many scholar-officials escaping from government service who built retirement homes with elaborate private gardens. Buddhism was suppressed in the north, and its influence moved south and across the Yangtze River where it was accepted. Gardens continued as miniaturizations of nature, with more symbolic and emotional meaning, especially for the scholar-officials who were withdrawing and retiring from government service. In this context, gardens as symbols of nature were strongly appreciated, especially for spiritual sustenance, emotional well-being and delight.

The Tang dynasty is widely accepted among scholars as poetry's "golden era" with many poets deeply religious. The growing literati and general population sought harmony with nature in their search for spiritual sustenance and delight. In the context of the second period, nature inspired garden-making with poetic and emotional explorations of nature's beauty. Hence, scholar gardens in the Tang period were designed to create poetic impact. Numerous temples and garden complexes were built during this period in addition to the donated properties. These were financed by merchants, as well as by nobles from the imperial court. Descriptions of residential villas and temples with their gardens were found in Tang poetry and literature; and these cultural artifacts confirm the importance of the garden for scholars and the general population as destinations (Feng 1992; Kewsick 1978; Johnston 1991).

Feng's third period, "Reasoning", overlaps with the Tang and later Song period. The "*shan-shui*, mountain-water" genre in poetry and landscape painting was popularized during this period with nature as the inspiration. It represented an intellectual appreciation beyond just demonstrating artful skills. Mountains, rivers and lakes were prominent compositional elements in the visual arts and were representative of an artist's reflection of the meaning of nature, inner reality and wholeness. As an accepted style for the Song imperial court, these landscape paintings and *shan-shui* poetry were perceived politically as representing a cultivated mind during one of China's stable periods.

This particular genre flourished when Hangzhou, currently in Zhejiang Province, was China's temporary capital during the Southern Song dynasty for nearly 150 years. The Song Imperial Painting Academy, the first official art academy in a re-unified China, was also established there. During this time, garden-making activities were considered extensions of China's classical arts. Some scholars argue the Song dynasty is the equivalent of Europe's Renaissance, especially with the flourishing of China's arts and culture, along with urbanization, and sophisticated accounting practices with the first use of government-issued bank notes. In Feng's Reasoning period, the importance of the "view" and strategy to deploy a series of garden scenes represented an intensification and appreciation for natural beauty.

Gardens in this period were structured around poems, where the "*jing*", view or garden scene, was considered the equivalent of a verse in a poem or the subject of a painting. One idyllic use of "*jing*" is represented in the iconic "Ten Scenes of West Lake, *Xi Hu Xijing*" located in Hangzhou and popularized during the Song era (Barmé 2012). West Lake's iconic spatial form and picturesque landscape incorporated an artificially created fresh water lake with islands and causeways surrounded by mountains on three sides; and it was duplicated throughout China over time. As its own cultural landscape, West Lake was the subject of poets, writers and artists, and a significant Chinese case study in its own right, and will be discussed in the fifth chapter within the context of the four late 20th century hybrid modern case study parks. However, the idea of West Lake's Ten Scenes is important for understanding the concept of "*jing*" in Feng's (1992) history of Chinese garden-making and Feng's third period. These ten scenes were meant to be enjoyed at different times of day and different seasons, and were marked by imperial gardens in the Song era. These also underscored the Song dynasty *shan-shui* literati tradition when painting, calligraphy, poetry and garden-making were intertwined.

The imperial gardens located at West Lake's Ten Scenes declined after the Song period. In the latter Qing dynasty, two emperors sought to reclaim West Lake's beauty by inscribing stone stelae with calligraphy and poetry at the sites of the classical ten scenes. In effect, the imperial gardens at these ten scenes represent a system of outdoor open green spaces linked together to represent the picturesque landscape of West Lake, an early innovation of landscape design in China and the world. Hangzhou Administration of Gardens and Cultural Heritage's (2011) nomination file for World Heritage recognition by UNESCO notes: "West Lake's ten poetically named scenes embody idealized classical landscapes that manifest the perfect fusion between man and nature." When compared to West Lake and the larger imperial parks, the smaller residential gardens designed for and by retired scholar-officials were designed as places where a musical instrument could be practiced or performed; poetry could be written; the game of "go" could be played; guests were entertained; and one could be immersed within the naturalistic scenery of artificial mountains, greenery and waterscapes. Sun (2008 interview) claimed that gardens were three-dimensional landscape paintings.

"Reasoning", Feng's (1992) third period of garden-making, depicts a time when "learning from nature" was intensified. Nature was a phenomenon to be explored and the garden with its compilation of a series of views provided a vehicle for this visually intensive, intellectual and sensory experience. The Chinese Picturesque garden dominated the visual picture plane with the buildings sublimated and imbedded in the landscape.

"Spirit", Feng's (1992) fourth period, a relatively short period, overlaps the previous period (Reasoning) and later period (Intention) and appears to formalize the "*shan-shui*", mountain-water tradition. The creation of artificial hills and water features in the garden were important design elements in the private scholar garden for this period. Garden meaning evolves into an immersive experience when garden-makers were completely absorbed by their naturalistic surroundings and gardens reflected the essence of nature with deep emotion. Water systems were integrated and involved artificially constructed water bodies: lakes, goldfish ponds and river-like courses, along with backdrops of artificially made hills articulate an idealized form of nature. In effect, this period accentuates the garden as a reflective agent conveying emotion. Scenes within the garden are composed with attention to the details of the five senses, aesthetics, and related poly-sensory and experiential qualities. In its most complex version, the garden was intended to be an experience that was sensual, philosophical, and intellectual. Jin (1992, p. 348) states "the experience of the garden went beyond the physical boundary of the garden into a world of imagination and allusion". The Chinese Picturesque and garden-making were at a level of enlightenment and greatest sophistication in terms of spiritual, deep emotional and intellectual pleasure. The garden in this period represents an idyllic representation of nature.

"Intention", Feng's (1992) fifth and final period, overlaps and spans part of the Yuan, Ming and Qing dynasties. The Ming and Qing dynasties are considered the peak of garden-making in Suzhou and southern China when numerous scholar-officials retired (forcibly or by choice) from the Ming or Qing imperial courts. Garden design is raised to the level of a professional craft and fully accepted as an extension of the four classical arts. Ji Cheng's 17th century book on the Craft of Gardens is evidence for this particular period of garden-making. Paving and path design are articulated around views of scenes that are revealed within the garden experience. As an artistic expression, some historians argue that the idea of nature is diluted and the garden becomes more superficial, as garden designers compete with each other to create unique spatial forms with artificial hills, the piling of stones and artful use of stones and rockery, water systems and features, vegetation, path design and garden architecture.

The meaning of nature evolved as a phenomenon that was re-organized in garden settings along with an intentionally choreographed garden experience – a composite of scenes that combined artificial, man-made materials with natural materials. Garden-making became an artistic expression and the designer's intentions were mindful of the effects created for the garden experience. Scholar-officials strived to make their gardens emblems of the literary tradition and showcases for their personal cultural accomplishments. Retired government officials designed gardens as contemplative places to continue their pursuit of the classical arts. As gardens for the merchant class became more common in the late Qing period, the garden existed primarily for visual entertainment. The meaning of the late Qing garden became less important and was a substantial simplification of earlier garden design approaches.

Chinese Picturesque garden design language

The scholar garden has not only become the Chinese garden archetype for the rest of the world, it also has a very special place in Chinese culture. The design conventions deployed in Chinese scholar gardens, especially the use of rockery, have come to be viewed as the essence of Chinese landscape design, a symbol of "Chineseness" to many of the Chinese public (Z. Bao 2009, pers. comm., 30 April). This extraordinary symbolic significance has helped to give the Chinese Picturesque a great deal of influence on contemporary designed landscapes in modern China. The design principles and spatial form are the source of China's most enduring garden design language and vocabulary. An understanding of this garden form is critical to interpreting contemporary practices of landscape architecture in China, especially in the late 20th century. The following narrative synthesizes a basic understanding of the spatial forms, design principles, grammar and vocabulary of the Chinese Picturesque genre represented in the private residential scholar-official's garden.

The consensus among historians is that the private scholar garden flourished in urban areas south of the Yangtze River in the current Jiangsu and Zhejiang provinces, historically known as the *Jiangnan* region, (literally south, *nan*, of the *Changjiang* or Yangtze River). Hundreds of scholar gardens existed in Suzhou, a canal town and popular retirement location, which was then known as *Pingjiang* (Keswick 1988; Johnson 1991; Graham 1938; Valder 2002). Valder (2002, p. 244) notes that the prosperity of the late Ming created "an unprecedented mania for building gardens among the retired officials, merchants and gentry". The late Ming dynasty was another period when China was part of the world economy; and their cultural artifacts like porcelain ceramics and silk textiles were exported around the world and consumed by western Europeans and the British. During the Qing empire, successful merchants sometimes hired professional garden designers to create classical and scholar-like gardens for them. These were known as merchant gardens, *shang ren yuanlin*. By that time, professional garden designers were utilizing the manual for garden design written by Ming dynasty gardener, Ji Cheng, as a reference.

In general, the design approach was based on *yi jing*, a Daoist metaphysical concept that sought to stimulate the garden experience of the visitor or owner through the mingling of sensory experience, ideas, thoughts and emotions with objective life and garden scenery (Feng 1992; Zou 2002). This was an aesthetic approach used in poetry and literature where the poet or designer would deploy "scenery manipulation, *jing se*", as a tactic to produce the desired experience. Garden meaning and *yi jing* changed and evolved over time with continuity of the basic elements of the Daoist concept of man's unity or harmony with nature.

The temporal dimension was a critical component for Chinese Picturesque design grammar. Garden designers considered the passage of time where the sensual experience of the four seasons was vital to the outdoor designed landscape. Plants were carefully selected to display seasonality as well as to enhance the auditory and olfactory senses. The Chinese tradition of moon-gazing also became a key element portrayed in garden design. In Zou's (2002, p. 293) words, "the obsession with moonlight was widespread in Chinese gardens". As the classical garden evolved into a simpler version during the decline of the Ming dynasty, its meaning shifted from an existential experience to a visual experience of gardens (Jin 1998; Zou 2002). However, the importance of this type of visual experience and symbolism from the *shan-shui* tradition continued through the Qing dynasty and into the 20th and 21st centuries.

In addition to scenery manipulation, the following key design characteristics and principles have been widely accepted for the formal expression of the Chinese Picturesque garden. Generally, these gardens were designed to be experienced on the basis of one or two ways: 1) visualization and 2) movement (Chen 1985; Keswick

1978; Johnston 1991; Jin 1998). Chen (1985, p. 41) describes one type as "in-position viewing" and the other "in-motion viewing". Johnston (1991, p. 75) describes these gardens as "those in which the garden is viewed from certain fixed positions and those where dominant impressions are gained while one is walking through the garden". The first garden form is based on the viewer's position at any location within the garden and is designed for passive experience. The second type is based on the viewer's pedestrian movement and mobility throughout the garden and is designed for a more active experience. Examples of these garden types are in Suzhou, Jiangsu Province, the smaller scale Master of the Nets Garden, *Wangshi Yuan*, which covers over one acre and has been replicated at New York's Metropolitan Museum of Art, and the larger Humble Administrator's Garden, *Zhuozheng Yuan*, covering thirteen acres or 5.2 hectares (Chen 1985; Johnston 1991). Although the design of scholar gardens was never meant to be made formulaic (Jencks 1978), scholars generally accept garden design principles and distinctive ways the garden deals with objects, space, and pedestrian mobility (Johnston 1991; Keswick 1978; Jin 1998; Zou 2002; Chen 1985).

The primary design elements or design vocabulary employed in Chinese Picturesque gardens include water, rockery, plants and architecture. These were linked through a pedestrian path design and circulation system that would "twist and turn" (Chen 1985). Residential buildings and related rooms were incorporated into the garden design and scenery. Some Chinese scholars argue that non-Chinese cannot fully grasp classical Chinese gardens without understanding the calligraphy, poetry and other textual ancestral narratives exhibited within the residential building and garden complex. However, many non-Chinese scholars and design historians agree that sensory garden experiences in themselves involve the appreciation of works of art, no matter the origin or provenance of the work, and the need to understand the textual artifacts may not degrade the overall garden experience or appreciation.

Certain key elements of basic architecture within the enclosed scholar gardens include the hall, *tang*; moongate, *yuegongmen*; pavilion, *ting*; a building that overlooks scenery, *lou*; covered walkway, *lang*; bridge, *qiao*; waterside pavilion, *xie*; and landboat, *fang* (Johnston 1991; Liu 1993). Rockery and stone symbolized mountains and other natural elements and were key critical three-dimensional design elements. These were crafted to form artificial mountains or rockery, *jia shan*, and also arranged as focal points or visual features in the garden. For Chinese garden experts, rock-piling, *die shi*, or creating artificial mountains is one of the most important design principles (Liu 1993).

The use of water, *shui*, was central to the overall garden design. As the dual companion or opposite (*yin*, female) to mountain, *shan*, or *yang* (masculine), water was fundamental to the mountain-water (*shan-shui*) landscape painting and poetry aesthetic. It provided desired reflections for moon-gazing and helped to create the illusion that small-scale gardens were larger. In both the "in-position" and "in-motion" viewing garden, water is the primary organizing feature, one of the most important scenes and a major visual focal point (Chen 1985; Liu 1993).

Plants, *zhi wu*, were also principal design elements for defining garden scenes. Plants were used to frame the view and enhance the view. They were also located to create focal points and signal seasonal changes. Plants were used individually rather than in a massing of plants typically found in western landscape design. In addition to emphasizing the visual experience in the garden, plants were selected and spatially arranged to enhance auditory and olfactory sensations. For example, classical Chinese literature described the sound of wind rustling through leaves of banana trees, as well as the fragrance of plants in the garden (Chen 1985; Keswick 1978; Zou 2002; Jin 1998). The full range of plant types were deployed in gardens: fragrant flowers, fruits, evergreen and deciduous trees, climbing plants, bamboo, herbs and water-based plants. Individual plant species carried important symbolism and its foremost purpose was to create an imaginary natural world. Individual plants were also used as symbols in poetry, literature and garden design to reflect a scholar's life (Valder 1999). For example, bamboo represents strength of character and pine trees represent wisdom. Buildings in gardens were typically named after plants (Liu 1993).

The treatment or structuring of "space" in gardens involved careful composition of the scene, *jing se*. Since many of the scholar gardens covered small tracts of land, space and visualization of the garden scenery were often manipulated. For example, carefully articulated visual and painterly devices and the intentional manipulation of a garden scene or view could create the illusion of making a small space seem larger. For example, the technique of "borrowing scenery", *jiejing*, was an important visual device used when garden design was at its peak. One of the garden views in the Humble Administrator's Garden incorporates an architectural feature outside the boundaries of the property but within the garden's sight line (Figure 3.4). In effect, borrowing the scene of this architectural element and backdrop expands the perception of space

Figure 3.4 Humble Administrator's Garden borrows the scene of the pagoda located outside the property, photograph by author

through visual and scenery manipulation (Jin 1998; Zou 2002). "Framing the scenery or view, *kuangjing*", was another device and could also involve walking through a gate, *men*. In this type of experience, a garden scene may exist beyond a gate (could be a moon-shaped gate or other articulated gate form) and after pausing to walk through the gate, the perception of the space in the next scene could also feel larger.

The set of scenes represented in a garden was designed from the ground-level pedestrian view rather than a bird's eye or aerial view. Visualization was carefully considered and, as Zou (2002, p. 300) stresses, "this view point was constructed as if looking at a painting". As noted earlier, the series of scenes within a garden was linked to poetry. The debate about whether a poem informed the scenes of a particular garden or a specific poem was written about a garden after it was designed remains unresolved today. Regardless of which came first, the traditional garden is composed of a series of scenes with each scene assigned a verse of poetry. The use of ten scenes in the garden design's composition is attributed to the ten classical scenes at the original West Lake in Hangzhou; and typically, these garden scenes were revealed to the visitor or owner as he or she experienced the garden.

Movement and mobility through the garden were carefully structured by combining the pedestrian path circulation system with garden scenes that divided the garden asymmetrically. This movement system utilized a combination of "naturalistic" or non-linear paths integrated with linear paths. The straight or linear path was associated with bridges, covered walkways, gates, doorways, and corridors. The curved path was designed to create links between the various garden scenes. Chen (1985) points out the experience of the garden as three dimensional and notes the importance of using curving paths in a manner complementary to straight ones. The curving path system surrounds the visitor with pleasant scenery and makes the route seem longer and more interesting. The use of curved pedestrian paths that "twist and turn" revealed an orchestrated set of scenes. Drawing out the pedestrian experience in this way provided the illusion of a larger garden and a perceived miniaturized natural world.

In addition to the horizontal path alignment and their spatial forms, designers introduced vertical elevation changes. For example, some gardens contain a covered corridor along the lake's edge with its path undulating up and down vertically in a wavelike pattern to create the feeling of water movement. In other garden scenes, ascending paths may meander through artificial mountains and lead to a pavilion designed to overlook the garden. This is intended to create the feeling of solitude on an imaginary mountain or the feeling of contemplation believed to occur at the end of a pilgrimage.

The device of "contrast and foil" is considered another important structuring principle for scholar garden design and captures the spirit of *yin-yang* dualism (Liu 1993). The importance of contrast in organizing the lay-out of scenes includes "openness and brightness" versus "tortuousness and gloom" (Liu 1993, p. 16). The use of foils creates a sense of visual hierarchy within the scene. For example, objects could be spatially organized within scenes as major visual features that could be distinguished from a minor object. This serves as a type of visual trickery that might involve the insertion of a proportionately small object within a scene to make the spatial experience appear larger than it is.

Pedestrian paths were intimate in scale with widths no wider than accommodating two people standing side by side and often narrower for use by an individual. The paths were typically made using various indigenous stone and brick materials. Paving patterns were designed according to the movement intended by the garden designer.

Intricate mosaic patterns, usually made of stone, marked focal points in the garden and changes in the garden scenery, as well as defined or emphasized a special garden scene. The use of bridges added to the sensory garden experience. These physically linked artificial islands and created places for day-dreaming, viewing the waterscape, and experiencing the surrounding scenery, especially during the four seasons.

Essentially, the Chinese Picturesque design language emulates the *shan-shui* mountain-water aesthetic employed in classical landscape painting and poetry or the imperial literati arts. It was a cultural product of the elite and its meaning was originally embedded in the Daoist interpretation of nature as a dominant force that man had to adjust to. In the garden's formal design expression, it was a painterly series of spatially articulated scenes representative of a poetic, romantic and imaginary nature. Keswick (1988) refers to this as "magical naturalism". The design elements of the Chinese Picturesque garden included water, rockery, plants, articulated pedestrian path systems, islands and architecture. The grammar for the classical garden involved a variety of ways to manipulate garden scenes, expanding the perception of space, dividing the garden space in asymmetrical forms, and creating places or observation points for contemplation, making music, painting and writing poetry, as well as places for gathering and socializing (Liu 1993).

The symbolism and conventions of the Chinese Picturesque style are commonly used to create a distinctively "Chinese" identity in contemporary designed landscapes in China. In some cases, neither the garden or park site or their designs have any concrete link to the Chinese Picturesque tradition or historic Jiangnan region. The language of the Chinese Picturesque genre has been removed from its physical and historical context. However, the traditional Chinese garden retains its meaning as a key symbol of Confucian elite or literati culture and continues to represent an abstract Chinese identity to the people of China. The melding of cultural identity and these Chinese Picturesque elements with the influence of recent international trends in design, as well as the "meaning" of sites, has helped to inform the design language of hybrid modernity represented in post-Mao China's late 20th century purpose-built parks.

References

Barmé, G. (2012) A Chronology of West Lake and Hangzhou, *China Heritage Quarterly*, 28. http://www.chinaheritagequarterly.org/features.php?searchterm=028_chrono.inc&issue=028

Birrell, A. (1999) *The Classic of Mountains and Seas*. London, UK, Penguin Publishing Group.

Bosker, B. (2013) *Original Copies: Architectural Mimicry in Contemporary China*, Honolulu, HI, University of Hawaii Press.

Chen, C. (1985) *On Chinese Gardens*. Shanghai, China, Tongji University Press.

Chen, X. & Wu, J. (2009) Sustainable Landscape Architecture: Implications of the Chinese Philosophy of "Unity of Man with Nature" and beyond, *Landscape Ecology*, 24(8), 1015–1026.

Clunas, C. (1997) *Art in China*. Oxford, UK, Oxford University Press.

Fairbank, J. K. (ed.) (1957) *Chinese Thought and Institutions*. Chicago, IL, University of Chicago Press.

Feng, J. A. (1992) Mutual Nutrition of Man and Nature: An Outline of the Comparative History of Landscape Architecture, *Spazio e Societe*, 57, 62–77.

Graham, D. (1938) *Chinese Gardens: Gardens of the Contemporary Scene: An Account of Their Design and Symbolism*. London, UK, Harrap.

Hardie, A. (1988) translation of Ji Cheng's, 1631, *The Craft of Gardens*. New Haven, CT, Yale University Press.

Hunt, J. D. (2002) *The Picturesque Garden in Europe*. New York, NY, Thames and Hudson.

Jencks, C. (1978) Meanings of the Chinese Garden. In: *The Chinese Garden: History, Art, and Architecture*, Keswick, M. (ed.). London, UK, Academy Editions.

Jin, F. (1998) Jing, the Concept of Scenery in Texts on the Traditional Chinese Garden: An Initial Exploration, *Studies in the History of Gardens and Designed Landscapes*, 18(4), 339–365.

Johnston, R. S. (1991) *Scholar Gardens of China: A Study and Analysis of the Spatial Design of the Chinese Private Garden*. Cambridge, UK, Cambridge University Press.

Keswick, M. (1978) *The Chinese Garden: History, Art and Architecture*. London, UK, Academy.

Keswick, M. (1988) Foreword in, Cheng, J. transl., *The Craft of Gardens*. New Haven, CT, Yale University.

Knapp, R. G. (1989) *China's Vernacular Architecture: House Form and Culture*, Honolulu, HI, University of Hawaii Press.

Li, Z. (2010) *The Chinese Aesthetic Tradition*, translated by Samei, M. Honolulu, HI, University of Hawaii Press.

Liu, D. (1993) *Classical Chinese Gardens of Suzhou*, translated by Chen, L. New York, NY, McGraw-Hill.

Siren, O. (1949) *Gardens of China*. New York, NY, The Ronald Press Company.

Steinitz, Carl (2008) Landscape Planning: A Brief History of Influential Ideas, *Journal of Landscape Architecture*, 3(1), 68–74.

Sun, X. (1984) Let a Hundred Flowers Blossom!: Classical Gardens in China. In: *Cultures/China, Past and Present*, 34/35: 51–61, Leander, B. (ed.). Paris, France, UNESCO.

Tan, W. L. (2011) Communal Worship and Festivals in Chinese Villages. In: *Chinese Religious Life*, Palmer, D., Shive, G., & Wickeri, P. (eds.). New York, NY, Oxford University Press.

Tang, Yijie (2015) *Buddhism, Daoism, Christianity and Chinese Cultures*. Berlin, Germany, Spring-Verlag.

Tao, Q. (1965) Peach Blossom Spring. In: *Anthology of Chinese Literature, Volume I: From Early Times to the Fourteenth Century*, Birch, C. (ed.). New York, NY, Grove Press.

Valder, P. (1999) *Garden Plants of China*. Portland, Oregon, Timber Press.

Valder, P. (2002) *Gardens in China*. Portland, OR, Timber Press.

Wang, L. (1999) Tourism and Spatial Change 1911–1927. In: *Remaking the Chinese City*, Esherick, J. W. (ed.). Honolulu, HI, University of Hawaii.

Weller, R. P. (2011) Chinese Cosmology and the Environment. In: *Chinese Religious Life*, Palmer, D., Shive, G. & Wickeri, P. (eds.). Oxford, UK, Oxford University Press.

Wright, A. F. (1977) The Cosmology of the Chinese City. In: *The City in Late Imperial China*, Skinner, G. W. (ed.). Stanford, CA, Stanford University Press

Yu, K. & Del Tredici, P. (1993) Infinity in a Bottle Gourd: Understanding the Chinese Garden, *Arnoldia*, 53(1), 2–7.

Yu, K. & Padua, M. (2007) China's Cosmetic Cities: Urban fever and superficiality, *Landscape Research*, 32(2), 225–249.

Zhao, S. (2002) Town and Country Representation as Seen in Temple Fairs. In: *Town and Country in China: Identity and Perception*, Faure, D. & Tao, L. (eds.). New York, NY, Palgrave.

Zou, H. (2002) The Jing of a Perspective Garden, *Studies in the History of Gardens and Designed Landscapes*, 22(4), 293–326.

4 Looking "inside" and "outside" post-Mao China

Looking "inside" and "outside" late 20th century China navigates post-Mao cultural development and informs hybrid modernity. In this light, this chapter examines China's cultural aftermath after Mao Zedong's death and late 1970s launch of Deng Xiaoping's policy of "reform and opening-up to the world, *gaige kaifang*". It reveals China's cultural awakening following Mao's period of global isolation, decade-long Cultural Revolution and the Communist Party of China's (CPC) abandonment of utopian totalitarianism. It touches on Mao's "official" CPC ideology, as well as unofficial movements, trends and themes expressed by China's writers and artists during the "open-ness" of Deng's New Era, *Xin Shiqi*. The post-Mao "spirit of an epoch, *shidai jingshen*" (Gao 2011) influenced the attitudes of artists and writers and in turn allied professionals working in the built environment, particularly landscape architects. Early development stages of the profession of landscape architecture and related design education are also touched on. The chapter sheds some light on China's cultural milieu at the turn of the 21st century as their membership to the intergovernmental World Trade Organization was formalized and their bid for the 2008 Beijing Olympics was successful. To understand the western world China opened up to, the chapter touches on "outside" China – cultural trends and international influences from the postmodern western world, particularly given the bombardment of a plethora of western publications and media on China's emergent post-socialist and late 20th century society. A review of the cultural context "inside" and "outside" late 20th century China, including international trends in landscape architecture, informs the transdisciplinary and discursive construction of hybrid modernity – a fusion of local and international influences intertwined with nation-building and local identity. As Gao (1999, p. 17) notes, "The Chinese consciousness of modernity has only recently begun to be transformed from a self-focused to an interactive one".

The late 1970s and early 1980s are widely accepted as China's "golden era" and cultural renaissance. As Deng's notion of the market economy or "socialism with Chinese characteristics, *zhonggou tese shehuizhuyi*", took hold, questions of independent creative expression, cultural identity and attitudes shifted as rapidly as the pace of China's late 20th century hyper-urbanization. China's artists and writers, whose independent creative expressions were largely repressed during the Mao era, were the thought leaders and cultural pioneers as Deng's reforms opened China to the postmodern world.

Navigating cultural development and the arts in China since the turn of the 20th century is highly complex. While there is not sufficient space in this chapter, or book for that matter, it is widely accepted that transformations in China's arts and literature over the course of the 20th century were a result of two major movements: 1) May

Fourth New Cultural Movement, *Wusi Xinwenhua Yudong*, circa mid-1910s–1920s; and 2) 1980s "Cultural Fever" *Wenhuare*, also known as the New Enlightenment movement, *Xin Qimeng yundong* (Li 1993; Andrews 1994; Sullivan 1999; Gao 2011). As suggested earlier in this book, China's first decades in the 20th century were a time when political leaders, intellectuals, artists and the literati grappled with cultural identity, modernization, modernity and a new China when it emerged from thousands of years of dynastic rule, and a largely feudal society. This involved wide-ranging and impassioned debates on the "cultural confrontation between East and West", *dongxi wenhua lunzhan* (Wang 2001; Dirlik 2002; Gao 2011). Often this debate occurred at academic conferences or political gatherings, and occurred in three stages: 1) analyses of the similarities and differences between China and the West; 2) comparisons of the respective worthiness and shortcomings of Chinese and Western cultures; and 3) deliberations on the future of Chinese and Western cultures.

This same debate took place soon after Deng's reforms when artists struggled with the dichotomy of anti-traditionalism and traditionalism, especially with many wounded and oppressed from public humiliation, torture and living underground during the ten-year dark period of Mao's Cultural Revolution. Mao's modern vision of a utopian socialist state was never realized in the face of his failed economic policies and the government's fractured politics. The post-Mao notion of whether or not Chinese culture required modernization and the meaning that modernism could have for contemporary art within the broader analysis of Western art was part of the early 1980s Cultural Fever, *Wenhuare* – a complex cultural debate that embraced western thinking, particularly early French Enlightenment and Anglo-American liberalism – a society based on individualism, democracy and capitalism, not socialism (Wang 2001; Zhang 1997). This largely intellectual debate was in opposition to the central government's framework that advocated socialism through economic development. Before delving into the post-Mao cultural context, a review of Mao era arts and culture is critical, especially as it informs the complexity of China's late 20th century.

A glance back at the arts and culture in the Mao era

Prior to the Communist Revolution and 1949 establishment of the People's Republic of China and subsequent chaotic effort to build a socialist nation, Mao and China's left-wing elite lived in Yan'an (also referred to as Yenan in the literature), Shaanxi Province. As noted in an earlier chapter, China's Republican government failed to create a unified nation wherein it devolved into a milieu of generalized violence, warlordism and banditry, civil and imperialist war, randomness and unpredictability. Political tensions within the Nationalist Party spurred a communist faction that eventually led to a guerrilla war between their two respective armies. The so-called Communist Red Army (later renamed the People's Liberation Army) and Mao were driven from their southern base in Jiangxi Province in 1934 by the Nationalist army. This military retreat involved the brutal "Long March" or 6000-mile (10,000-kilometer) trek to Yan'an, a remote isolated mountain village in northwestern China that served as Mao's base of operations through the late 1940s.

Over 40,000 people arrived in Yan'an after the Long March in October 1935 (Hung 2011). Soon after Japan's invasion in 1937 and extended occupation, many of China's cosmopolitan intellectuals, artists and leaders of the earlier May Fourth and New Culture movements, circa 1916 to the mid-1920s, fled to Yan'an. By the mid-1940s, Yan'an became known as a mecca for communism and resistance against the Nationalist army and Japanese occupation with its population at over 2.5 million (Seldon 1995; Apter 1993). In support of their new base, Mao, communist party leader and his so-called party cadres established numerous schools, research institutes and training programs unique to their ideology, including the Lu Xun Academy of Fine Arts, Yan'an University, Central Party School and Women's University, to name a few (Zhong & Cui 2015).

The physical isolation from the Japanese occupation created a sheltered environment for Mao and his cadres to develop the CPC's culture-based ideology for China's new socialist society. Mao re-interpreted and transformed the Marxist notion of the proletariat derived from an advanced industrialized society (western European) to a socialist society that was largely rural-based and populated by the "masses": workers, peasants and soldiers, *gong ning bing*, as well as party cadres (Zhong & Cui 2015). Mao believed China's rural-based majority population was illiterate, uneducated and impoverished and hence could readily be politically influenced and organized to learn the "state" or CPC ideology.

Culture was perceived as the core of Mao's vision for a utopian socialist society. In this context, democracy, freedom from western colonial influences, nativism, individualism in the arts and culture were viewed as ways to create a modern nation by the leaders of the May Fourth and New Cultural movements. Mao and the CPC framed the utility of culture as the ultimate method for creating a new socialist society. In this Marxist-based narrative, Mao attempted to transform the meaning of Chinese identity from an elitist, cosmopolitan intellectual and individual point of view to one that focused on the masses (Hung 2011). During the Yan'an period, intellectuals and party leaders debated the socio-political ideology for Chinese society. Eventually, the May Fourth leaders' perspective was over-ruled and silenced by Mao's vision of a peasant-based socialist society – steering away from cosmopolitan individual elitism or petty bourgeoisie perspectives toward a collective individualism among the masses (Cheek 1984; Hung 2011).

The core of Mao's cultural ideology for shaping China's new socialist society was revealed in his 1942 Talks at the Yan'an Forum on Literature and Art, *Yanan wenyi zuotanhui*. While scholarly works have interpreted the English translation of these talks, literature written since Deng's reforms reveals more light on Mao's deployment of culture in his nation-building efforts. The widely accepted interpretation is Mao's assertion that art and literature should serve politics and reflect the experiences of workers, peasants and soldiers – not "art for arts' sake".

Before highlighting culture through the lens of Mao's perspective, it's important to note that Mao's personal background was unusual among his peers in that he did not study overseas. He was not part of China's "returned students, *liuzhuesheng*", selected youth in the late 19th and early 20th century decades who were sent to study in Japan, France and other western nations. His personal experiences reflected China's 20th century development and eventually shaped his leadership. His brief biography indicates Mao was born and raised to family farmers in the Hunan Province circa 1893; at age fourteen he rebelled against an arranged marriage, studied at a local secondary school, and subsequently joined Sun Yat'sen's Revolution as a military soldier against the Qing government (Devin 1997). After further studies, including the Confucian classics in Changsha, Hunan Province, Mao was certified as a school teacher. He uprooted after his mother's death and found his way to Beijing where he found a job as an assistant librarian at Peking University, *Beida* (Hung 2011). Mao's arrival in Beijing coincided with the dawn of the New Culture and May Fourth movements when he was introduced to Marxism through university lectures and self-study. Immersed in Beijing's radical student activities, Mao engaged with future leaders of the CPC – all eventually found their way to Yan'an by the late 1930s (Hung 2011).

At the 1942 Yan'an Forum on Literature and Art, Mao framed his concept of culture and socialist art as "national forms", *minzu xingshi*: new representations of Chinese traditional art infused with political messages that could be easily understood and appreciated by the masses (Andrews 1994; McDougall 1980; Hung 2011). It was within this spirit that Mao appropriated from popular China's folk tradition of "storytelling" as a way of "teaching" the masses (Apter 1993). Later, Mao launched various political campaigns and activities as methods for governing the People's Republic of China. Over the course of these campaigns, propaganda art emerged as Mao's primary tool to communicate new policies, generate a new nation-building spirit and teach the masses about China's new socialist society. Highlights of Mao's deployment of national forms of socialist art are summarized below.

Soon after the 1st of Oct 1949 establishment of the People's Republic of China, Mao deployed the *"nianhua"*, a folk art tradition involving inexpensive printing techniques, as the inaugural cultural propaganda tool (Hung 2011). He sought to unify the nation by celebrating the coming lunar new year through a new style of *nianhua*, a widely accepted traditional annual family practice. The colorful and festive *nianhua* (imagery included deities and other folklore) typically adorned entryways to residential dwellings during the family-oriented Spring Festival new year celebration that spanned several days. The proposed 1950 *nianhua* aligned with Mao's assertions on art and culture at his 1942 Yan'an talks – that art would serve the masses and depict the everyday life of workers, peasants and soldiers as positive, heroic and uplifting (Andrews 1994; Apter 1993).

Mao's cultural ideology was further reinforced by Zhou Enlai (1898–1976), one of the long-serving and significant CPC leaders. During his July 1949 Beijing lecture at the All China Congress of Literary and Art Workers, *zhonguo quanguo wenxue yishu gongzuozhe daibiao dahui*, Zhou emphasized: 1) unification of all China's literary and art workers was essential and inclusive of those in regions controlled by the People's Liberation Army (PLA) and areas controlled by the Nationalists; 2) artists were to serve the people, especially the workers, peasants and soldiers; 3) popularization was to take precedence over raising of standards; 4) old literature and art were to be remolded so that content and form were unified; and 5) artists and art leaders must avoid particularism but instead consider the needs of the whole nation in their art (Andrews 1994).

Near the end of 1949, the newly established Ministry of Culture issued a directive to produce a re-interpreted "new" *nianhua* intended to celebrate China's new lunar year or 1950 Spring Festival. The text of the Ministry of Culture's directive was published in the official newspaper, *People's Daily*: "*Nianhua* should emphasize laboring people's new, happy, and hard-fought lives and their appearance of health and heroism. In art we must fully utilize folk styles and strive to capture the customs of the masses" (Flath 2004, p 127). These prints shifted away from the black and white woodcuts created in Yan'an to imagery representing "red, bright and shining", *hong guang liang* (Andrews 1994). As the first culture-based directive launched by the central government, it set a precedent and tone for the official production of art as a political tool throughout the Mao period, so-called propaganda art. The style of this new *nianhua*, then a type of graphic propaganda poster, was also influenced by Socialist Realism, the Soviet sanctioned nationalist art (Flath 2004; Meserve & Meserve 1992).

Imagery in Mao's propaganda art typically glorified efforts of the peasants, workers and soldiers. Later, as Mao felt his political position threatened, propaganda art depicted him in a paternal position and as a dominant figure over smiling peasants, workers, and soldiers. As noted in an earlier chapter, Mao transformed all of China's institutions to reflect Soviet ways. Other forms of visual art and graphic representation like cartoons, *manhua*; traditional brush painting, *guohua;* and oil painting, *youhua*, were morphed into Mao's form of Socialist Realism and nationalization of art, *yishu minzuhua* (Li 1993). Independent works created individually by China's artists, writers and intellectuals were actively oppressed by the CPC and Mao's totalitarianism approach to building a socialist nation.

In an attempt to actively engage China's intellectual community in 1956, who by then were estranged by the government's harsh ideological offensive against them, Mao encouraged free speech and the use of "Big Character, *dazibao*" posters, through his slogan "let a hundred flowers bloom, let a hundred schools of thought contend", announced at a 2nd of May speech (Sheng 1990). This so-called Hundred Flowers campaign, *Baihua yundong*, was considered unique given it encouraged both "independent thinking and free discussion" (blooming and contending) and invited criticism of the CPC for their mistakes in previous interactions with intellectuals (Goldman 1981). In Mao's Yan'an period, *dazibao* (hand-written large Chinese characters on inexpensive paper and posted on buildings and in public places), so-called "big character" posters, were a spontaneous way to express political dissent and an accepted everyday practice there.

Within a year of Mao's 1956 speech, students at Peking University displayed big character posters representing a variety of expressions: poems, cartoons, satirical essays and lengthy formal essays critical of China's government or so-called "state" (Sheng 1990). Additionally, numerous written works in official and unofficial publications were forcefully critical of the CPC's bureaucracy, nepotism and corruption. Mao, terrified by the aggressive criticism, volume and bitterness expressed towards the government and CPC, terminated the Hundred Flowers campaign in June 1957, and within weeks of its termination, Mao launched the "Anti-rightist campaign", *Fan Youpai yundong*, (Sheng 1990). This campaign targeted critics of the government who were deemed bourgeois and politically "right wing"; and hence, considered "enemies of the party". During this two-year campaign period, over 500,000 educated professionals, artists, writers and scientists were identified as "rightists" with thousands executed; and hundreds of thousands were exiled to the countryside "to rectify their thinking through labor" (Goldman 1981; Dillon 1998). Following this purge, big character posters and publications critical of Mao and the CPC vanished from the public eye. However, Mao observed firsthand the power of these posters to communicate with the masses; and he actively encouraged its use to support his leadership and attack his adversaries (Sheng 1990).

For years, hundreds of thousands of propaganda posters, *xuanchuanhua*, were produced in the collective work units and national publishing houses (Sheng 1990). These aligned with Mao's cultural principle of creating a national form of art that depicted workers, peasants and soldiers in heroic ways – "sublime, outstanding and perfect, *gao da quan*" (Meserve & Meserve 1992; Li 1993; Andrews 1994). The Hundred

Flowers and Anti-rightist campaigns are examples of the Mao era – a milieu of fear, confusion, chaos, contradiction and unpredictability when propaganda art manipulated the masses.

By the 1960s, poor economic growth and failed policies, including the Great Leap Forward, caused famine, poverty and widespread loss of life and especially represented Mao's failed efforts to build a utopian socialist nation. Major tensions, fractures and factions emerged then within the CPC. To maintain power within the CPC and re-activate his power base, Mao launched the 1966 Great Proletariat Cultural Revolution, also known as the Cultural Revolution (CR). Mao's goals were to rid the CPC faction of "enemies of the state"; renew the CPC with anti-capitalist ideology; and rejuvenate the notion of the proletariat and his rural political power base. He promoted the "spirit of revolt", *zaofan jingshen*, against capitalists or anti-proletarians, and incited extreme violence through his mobilization of Red Guards, young secondary school and university students (Meserve & Meserve 1992). Mao's revolutionary slogans and political campaigns like "Bomb the Capitalist Headquarters" and "Destroy the Four Olds and Cultivate the Four News", were represented in propaganda art. "Cultivate the Four News" in this key CR slogan was never defined and was eventually lost in the political message. Overall, these propaganda posters became known as "revolutionary realism", *gemingdi xianshi zhuyi*, and were intended to mobilize the masses during the CR (Meserve & Meserve 1992; Li 1993; Andrews 1994). By late 1968, the CR's extreme violence caused overall economic decline, high employment rates and very low economic activity in cities throughout China. With concerns for public safety, Mao launched another political campaign: "Down to the Countryside" or "Up to the mountains, down to the villages", *shangshan xiaxiang*, also known as the rustication movement (Spence 1991). Millions of educated urban youth, *zhishi qingnian*, were sent to experience rural life among peasant farmers with the intention to be re-educated. Propaganda posters were created during this rustication movement to mobilize China's masses.

The anarchy and chaos of the CR caused tremendous tension among the CPC leaders with two major factions emerging by the early 1970s. The so-called Gang of Four, *siren bang*, favored political mobilization, class struggle, anti-intellectualism, egalitarianism and xenophobia; and the second faction led by Zhou Enlai and Deng Xiaoping advocated for economic growth, stability, educational progress and pragmatic foreign policy (Dirlik 2002). Mao's death in 1976 and the CPC's Eleventh National Congress in 1977 officially terminated the CR, and incorporated a proclamation to "emancipate the mind" – liberating the nation from Mao's CR ideology. This spawned the movement to "Liberate Thinking, *xing qimeng yundong*" (Carrico 2017). 1977 marked the emergence of Deng Xiaoping as China's paramount leader along with the government's resumption of the University/College Entrance Examination re-opening higher education institutions throughout the nation (Sullivan 1988).

As Deng's reforms opened China to the world, members in the Gang of Four were publicly tried and found guilty of unjust persecution, harm and murder of thousands of people during the Cultural Revolution. At the same time, China's artists and writers initiated a healing process. It's important to note that many of late 20th century artists in China considered themselves intellectuals. In this role, they believed they were both socially and intellectually responsible for the cultural and spiritual well-being of society. The next chapter will bring these influences on late-20th century park design to light. China's late 20th century artists also wrote profusely about their work during their creative process and published these manifesto-like essays about post-Mao art in journals and weekly periodicals for debate in academic settings and private group meetings (Liao 1993; Li 1993; Gao 1999; Andrews 1994). Major themes and cultural trends in China's late 20th century are highlighted below.

Inside post-Mao China

The aftermath of the wholesale adoption of Soviet thinking and importation of experts mentioned earlier would carry into China's late 20th century post-Mao art world. During Mao's era, Soviet art educators were imported to teach students at academies and schools in China, and students were also sent to USSR's art academies where they learned classical painting techniques. These Soviet-trained artists returned to China's national academies of art, where they taught the concept of Socialist Realism – a government-sanctioned national aesthetic based on a Marxist-Leninist notion that obligated artists to reflect society through a political lens, particularly the CPC beliefs (Vaughan 1973). It's also important to note that an older generation of artists who survived the Cultural Revolution were foreign-educated at art institutions in Japan, France and other western locations where they were exposed to modern art in the first three decades of the 20th century. Prior to Mao's establishment of the People's Republic of China, many of these artists were searching for modernity and a cultural identity as a result of their disillusionment with traditional Confucian literati culture; and in their quest embraced western realism. As China's universities and art academies re-opened in the late 1970s, classical and western painting techniques were taught and another form of realism emerged (Li 1993).

Art historians, scholars and critics have broadly mapped the post-Mao period in two ways: 1) two major movements (the New Enlightenment movement, *Xinqi Meng yundong*, also known as Cultural Fever, *wenhuanre*; and the 1985 New Wave, *Bawu xinchao*); and 2) five chronological periods of development with overlapping trends and themes: 1979–1983; 1984–1986; 1987–1989; 1989–1994; and 1995 to the present time (Li 1993; Gao 1999). China's post-Mao art world is conventionally portrayed in terms of individual artists, groups of artists working in collaboration, regional schools, curated exhibitions, writings and publications, a variety of media including visual art, sculpture, installations and performances. Late 20th century artists practicing in China have also been temporally benchmarked by the Cultural Revolution: those who studied abroad before the Cultural Revolution; those who came of age during the violent and chaotic 1966–1976 decade; those born in the late 1960s who were the first generation to experience a cultural renaissance after Deng's reforms reopened China to the world; and those born in the late 1970s who had no firsthand experience of the Cultural Revolution.

While Deng framed China's new market economy, as socialism with Chinese characteristics, the production of "literature and art, *wenyi*" was still largely controlled by the state or central government. In this light, post-Mao late 20th century artists, writers and their works were framed through the lens of "official" (state-sanctioned via national policy, exhibitions or events sponsored by national art academies or other central government cultural institutions) and "unofficial" art – creative works ideologically in opposition to official art. Official designations of artists working "above" ground by central government evolved over time in the post-Mao era. As a carry-over from the Soviet-influenced Mao era, artists in the 1970s were formally recognized as part of a government-sanctioned or registered society, i.e. "painting society, *huahui*", and in the 1980s referred to as "art group, *qunti*", i.e. Xiamen Dada *qunti*.

As rampant urbanization or "urban fever" exploded in the early 1990s, artists who chose to remain in China claimed their own territories by colonizing villages or areas in the urban fringe – remnants of built fabric consisting of traditional courtyard dwellings slated for demolition and self-identified as informal or unofficial artists' village, *huajiacun*, or art village, *yishucun*. Some "squatted" in un-occupied residential buildings and lived rent-free; and other artists paid very low rent. These art villages evolved as places were artists lived, established studios for their art practice, and in some instances set up galleries for displaying and selling their works (Liu et al. 2013). Following the June 4th 1989 Tiananmen Square massacre, these artists moved away from the public eye and worked underground primarily due to the government purge of dissidents and intellectuals.

By the mid to late 1990s, a handful of Chinese artists became part of the international art market for collectors and art curators. In turn, local Chinese institutions recognized the vitality of art as a cultural commodity for international consumption and engaged in "territorializing" with the formation of official "art districts, *yishuqu*". These emerged in urban fringe areas and typically involved the adaptive re-use of former industrial or factory buildings built in the Mao era and in some cases involved a direct link to academic art institutions. The first exemplar occurred soon after Beijing's Central Academy of Fine Art relocated to an industrial building complex in the northeastern edge of the city. Artists who may have been displaced from art colonies would soon set up studios in the same building complex along with commercial galleries in what is now known as 798 Factory, also known as Beijing's 798 Art District, *798 Yishuqu* (Gao 2011; Liu et al. 2013).

Art historians and cultural theorists concur the emergent ideology among Chinese artists in the 1970s, 80s and 90s was based on a multi-dimensional and interconnected set of factors: the oppression of individual expression, humiliation and tremendous suffering during the Cultural Revolution with Mao's totalitarian state, revolutionary realism and propaganda art; Deng's economic revolution and transformed CPC leadership; and shift from global isolation to the onslaught of western ideology via various modes of communication and dissemination. This included translations of published written works and creative artifacts; exhibitions and lectures by visiting western artists and scholars, audio-visual media such as western television programming, radio broadcast and film, overseas studies, and the internet and worldwide web by the late 1990s and turn of the 21st century. China's artists and thought leaders also developed a renewed interest in humanism and the meaning of "Chineseness".

At the same time, policies by China's single party government vacillated on the degree to which artists and society could be influenced by western ideas or "opened" to the world. These involved two critical anti-western political campaigns: 1983 Eliminate Spiritual Pollution, *qingchu jingshen wuran*; and 1986–1987 Anti-Bourgeois Liberalization, *fandui zhichangjie ziyouhua*. Both influenced or negated cultural innovation and caused many artists and writers to work "underground, *zhuanru dixia*" or relocate away from China and work in self-exile (Sullivan 1988; Gao 1999; Carrico 2017). Deng, China's first post-Mao paramount leader, connected economic reforms and modernization with culture when he declared the key to China's success was to respect knowledge and talented people.

Post-Mao art trends

The majority of works of art and official exhibitions in the 1979–1983 period depict realistic imagery of the physical and psychological suffering and tragedy of the ten-year-long Cultural Revolution, and related impoverished everyday life in the country-side. Historians and art critics have identified these official works by artists (primarily painters) trained in China's art academies as "Scar painting, *shanghen huihua*", an extension of "scar literature, *shanghen wenxue*", and considered part of the broader "New Realism, *Xinxianshizhuyi huihua*" painting genre. The realistic "scar" imagery expanded from the earlier genre of Socialist Realism and Mao's revolutionary realism. These works also represented social criticism and intellectual discussions on human-ism that confronted the societal rupture created by the Cultural Revolution. Works created in this first post-Mao period were also aligned with the official ideology of the 1977 Liberate Thinking campaign, *xing qimeng yundong*. Deng urged the Chinese people to "seek truth from facts" and "liberate their thinking" from the darkness, dis-ruption and stifling of independent thinking experienced in the Cultural Revolution.

Art historians have also highlighted the significance of the unofficial exhibition on 27 September 1979 by the Stars Group, *Xingxing Pai*, twenty-three independent art-ists who were considered amateur (not formally educated in China's art academies). It was organized independently (not through official government channels) by the artists whose intentions were to publicly celebrate their liberation from the tyranny of the Mao era with works of art expressing individual creative freedom. Unable to get the necessary authorization to exhibit their work in an official state-sanctioned venue, the artists displayed their work (generally derivative of western modern art genres like Impressionism, Surrealism and German Expressionism) on a fence in a public park adjacent to the official China Art Gallery (Li 1993), now known as the National Art Museum of China in Beijing. It was the first time since the Cultural Revolution when the general public had the opportunity to view works of art repre-senting western modern genres – visual art unlike Mao's revolutionary realism and propaganda art they were accustomed to.

Given the exhibition organizers did not secure the appropriate permits, the local police declared it illegal and began dismounting the works the following day. In imme-diate response to the police action, the Stars Group (the core were in their 20s) self-organized and held a public demonstration a few days later on October 1st, a major public holiday and the 30th anniversary of the founding of communist China. This commenced with a protest march at the Democracy Wall on Xidan Street (the site of the earlier winter 1978–1979 Democracy movement, *minzhu yundong*, or Beijing Spring) and ended with speeches on individual rights and freedom of expression for artists, writers and all people in front of Beijing's municipal headquarters (Berghuis 2012). Acknowledging their demands, the China Art Gallery authorized exhibition of their works from 24 August through 7 September 1980 after they registered as the "Stars Painters' Society" with the official Beijing Artists' Association; and over 300,000 people visited the exhibition (Gao 1999; Sullivan 1999).

The members of the Stars Group have been deemed pioneers by art historians and critics as China's first wave of avant-garde artists whose works did not follow the government-sanctioned realism genre. Their creative artifacts represented an authentic spirit of individual creative freedom and cultural awakening. They were considered the first generation of artists who self-identified as populist or part of the working class, not part of the elite class of artists who had formal art education (from domestic or international institutions) or had official affiliations with art academies. Furthermore, their October 1979 protest march is considered China's earliest example of performance art and socio-political activism (Berghuis 2012).

Artists practicing in the 1979–1983 period were influenced by "Cultural Fever, *wenhuare*" – a movement reminiscent of the May Fourth and New Cultural movements from earlier in the 20th century. Cultural historians have likened this early post-Mao period to the Weimar Republic in Germany – a time of political freedom and new cultural creativity (Dirlik 2002; Zhang 1997; Barmé 1999). Some of the artists and writers believed that Sun Yat-sen's philosophy on democracy would be revived (Zhang 1997). In fact, the flood of international media, translations of western publications coupled with the subsequent art installations and performance art or "Happenings" in public places created an "open" milieu where cultural discussions took place among all strata of Chinese society (Gao 1999; Zhang 1997; Davis 1995).

Outdoor public open spaces (streets, plazas, parks) in China's major cities during this early post-Mao period were territorialized as places for "Happenings" or performance art. In this context, groups of artists created urban spectacles where the "shock of the new" provoked authorities while stimulating thought among the general populace. In some ways, these artists initiated the Chinese discourse on the "public sphere" as interpreted by German philosopher Habermas – a space for public discourse and communication (Calhoun 1989).

Artists from this early post-Mao period saw themselves as political activists and social pioneers. They believed their work served as demonstrations on ways to liberate society through individual expression and learn from the West (Gao 1999; Zhang 1997). For many, this was an intense period of cultural reflection and cultural criticism as transformative as the May Fourth movement. On the other hand, the government's response to this early post-Mao period of liberalization of cultural values and individual independent creative expression involved the short-lived 1983 Eliminate Spiritual Pollution campaign. It represented the CPC's "struggle, *douzheng*" with western liberal values, peaceful evolution and cultural contamination; and specifically targeted sources of spiritual pollution: artists, journalists, pornographers, and political theorists. For example, nudity in performance art, paintings and sculpture, and editorial cartoons in newspapers were considered pornographic; and written articles and editorials in newspapers or academic journals that represented political dissent caused spiritual pollution.

As a spiritual cleansing process, the 1983 Eliminate Spiritual Pollution campaign would make way for the development and flourishing of China's "socialist spiritual civilization, *shehui zhuyi jingshen wenming*" (Carrico 2017). Li (1993, p. XIV) refers to the driving force of this early post-Mao period as the "Demand for Stylistic Freedom and the Return to Humanism". The political tensions and harsh criticism caused by this nearly month-long campaign enabled the central government to dismiss individuals from official positions. At the same time, some artists retreated from public life into self-exile and worked underground in China; and others relocated away from China. For example, the Stars Group voluntarily broke up and the majority of the original members emigrated to Japan, Europe and the USA. Ai Weiwei, an original artist in the Stars Group, now internationally known, then moved to New York City.

Artists practicing, studying and graduating from China's art academies in the 1984–1986 period were exposed to translations of the western canon of modern art, theory, criticism and philosophy. They are considered China's influential 1985 New Wave,

Bawu xinchao, artists whose work transcended China's state-sanctioned realism. Noted art historian and curator Gao (2011) has framed this group of artists as China's avant-garde, *qianwei yishu*, who experimented with western forms of modernism: Dada, surrealism, German expression, conceptual art and American Pop art. This three-year period is considered a time of cultural criticism, self-critique, idealism, heroism and protest (Li 1993; Gao 2011). 1985 New Wave artists wrote profusely including manifestos, or writings as part of their creative process, as well as for academic meetings, conferences and publications. The majority of these artists were educated in China's art academies after the Cultural Revolution where they learned western oil painting techniques. They were also unharmed directly by the 1983 Eliminate Spiritual Pollution campaign.

The 1985 New Wave artists were also heavily influenced by the 1985 exhibition of self-curated work by American artist Robert Rauschenberg at China's Art Gallery in Beijing. His work exposed China's artists and the general public to a variety of western modernism and media: pop art, sculpture, and installations of everyday found objects (Gao 1999). This further spurred conceptual art and inspired outdoor installations in public places. China's 1985 New Wave artists were also responding to western-style capitalism and consumerism symbolized by American fast food chains like McDonald's, Pizza Hut, and Kentucky Fried Chicken – imagery that dominated China's everyday urban environment.

For many, the 1984–1986 period was the peak of creative development of an entirely new aesthetic. It was considered a time of deep cultural reflection, *wenhuan fansi*, as transformative as the May Fourth movement (Gao 1999; Zhang 1997; Davis 1995). By late 1986, China's artists and students became disillusioned with CPC's conservativism and growing sense of corruption. Students responded with public protests and demonstrations by late 1986. In response, the central government launched another anti-western ideology political campaign known as the Anti-Bourgeois Liberalization, *fandui zhichangjie ziyouhua*. It involved another purge of student dissidents, and a form of government censorship that caused avant-garde artists to work underground or move overseas. Artists and intellectuals were considered members of the bourgeoisie who required censorship and eradication. This new policy effectively halted innovation in the visual arts, especially performance art and social engagement within the public sphere.

In the subsequent 1987–1989 period, artists responded to the Anti-Bourgeois Liberalization campaign with work that reflected the theme "back-to-the-roots" or "search for cultural roots", *wenhua xungen*. Those artists who practiced "above" ground considered Eastern philosophy and re-visited traditional literati art techniques (calligraphy, ink and watercolor) with works that represented the "New Literati" genre (Li 1993). Other artists reflected on the meaning of "Chineseness" or Chinese identity through conceptual explorations of traditional materials and forms of representation, for example, the scroll format utilized in traditional literati art.

Xu Bing's 1991 installation, "Book from the Sky", mixed media, created 1987–1991, is an example of a conceptual work that appropriates from traditional literati art (Figure 4.1). This immersive work consists of an orderly arrangement of hand-printed bound books that appear to represent classical Chinese texts; numerous pages of text inscribed with nonsensical Chinese characters in a Song dynasty font style that artisans used in the Ming dynasty; long scrolls inscribed with fake Chinese characters hanging from the ceiling; and the walls covered with scrolls also inscribed with fake Chinese characters (Gao 1999). The work appeared satirical to some and absurd to

Figure 4.1 Xu Bing, *Book from the Sky*, 1987–1991. Mixed media installation/hand-printed books and scrolls printed from blocks inscribed with "false" Chinese characters. Installation view at Elvehjem Museum of Art, Wisconsin, 1991, © Xu Bing Studio

many, particularly given the labor and number of years that went into creating the work.

However, Li (1993) noted the significance of the search and need for artists to create a new Chinese aesthetic – revealing new meanings of traditional cultural symbols emerging out of western modes of conceptual art. Li (1993) also pointed out the shortcomings of the works created during the "back-to-the-roots" period as predictable, formulaic and commercially oriented. One of China's prominent literary theorists argues the New Enlightenment movement reflected a naïve belief that capitalism and consumerism could spontaneously bring about democracy (Wang 2003). Any innovation by artists was stalled by the event that brought this period of the post-Mao era to an end – the 4th of June 1989 Tiananmen Square massacre.

Culturally, 1989 was a milestone year for China. The China Art Gallery in Beijing held "The China/Avant-Garde" exhibition (conceptualized in 1986) for two weeks in February with hundreds of works (variety of media: paintings, sculpture, conceptual and performance art, film and documentary photographs of performance art) by nearly 200 New Wave artists. The 1986–1987 Anti-Bourgeois Liberalization campaign was relaxed by then, and artists in this authorized and controversial exhibition tested China's legal limits (Li 1993; Gao 1999). China's New Wave artists were introduced to their nation through this exhibition and later the western art market at venues in Berlin, West Germany; Rotterdam, Netherlands; and Oxford, England. It represented the peak of idealism, experimental art and broader debate for the establishment of new art in China – to remake art theory, art criticism and art administration (Gao 1999). China's avant-garde artists made their way to the international scene in a more significant way via the exhibition entitled "China's New Art, Post 1989 with a Retrospective from 1979–89", held first in Hong Kong circa 1993 and traveled to cities in Australia and the United States.

The China/Avant-Garde February 1989 exhibition in Beijing represented China's *zeitgeist* and took place two months before the non-violent student-led Democracy movement, *minzhu yundong*, when students sought institutional reforms with demands for the right to vote and speak freely. It is widely known that the spirit of their non-violent protest and public demonstrations influenced other students and the general public throughout China. The students were mindful of the convergence of the 200th anniversary of the French Revolution, 70th anniversary of the May Fourth movement and 40th anniversary of the founding of the People's Republic of China. The Democracy movement eventually ended with the tragedy of the June 4th Tiananmen Square massacre of over 10,000 un-armed people by the PLA outfitted with guns and tanks or so-called armored personnel carriers. This inhumane massacre was witnessed worldwide through news coverage provided by foreign media, who were there to cover the first Sino-Soviet summit since 1959 that included Soviet leader, Mikhail Gorbachev, and the Asia Development Bank's inaugural meeting in China (Calhoun 1989).

The 4th of June Tiananmen Square massacre and subsequent purge of dissidents broke the spirit of idealism among China's students, artists and writers. Many left China and those who remained experienced national malaise. Artists then worked underground in a style that became known as "Cynical Realism", *wanshi xianshi zhuyi*, and "Political Pop", *Zhengzhi bopu yishu*. The cult of Mao, revolutionary realism and propaganda art were phenomena explored by these artists, as well as the meaning of being Chinese. Those who went into exile, so-called Chinese transnational artists, developed their careers in major art centers like New York, London, and Paris. Their work is now highly valued by art collectors.

By the mid-1990s, the impact of modernization and rapid urbanization on the environment became visibly and physically apparent to the Chinese people. Environmentalism emerged as a new theme in Chinese art. Betsy Damon, an American artist, organized unofficial events in Lhasa, Tibet and Chengdu, Sichuan Province to raise consciousness about China's polluted rivers in the mid-1990s. The latter consisted of numerous artists (primarily Chinese) engaged in performance art and installation works in public places that raised environmental consciousness, especially about Chengdu's pollution of the Fu-Nan river system. Various creative activities were covered by media and local television catching the attention of Zhang Zhi Hai, a local administrator and director of the Funan Comprehensive Revitalization project. Zhang sought out Damon and their discussions set the basis for Chengdu's Living Water Park, a purpose-built project and the first of the hybrid modern case study parks discussed in the next chapter.

Themes of nostalgia for an earlier China also emerged in the visual arts, especially in photography, video and film, as well as traditional forms of expression in painting and sculpture. This was informed by the loss of historic buildings in urban centers and the merging of villages into expanded districts of cities (Gao 1999). The booming market economy of the 1990s spawned a new generation of artists who were products of China's elite art academies. The majority of these young artists were born after the Cultural Revolution and came of age in the mid-1990s post-Deng period. The entirety of their experiences took place in a society of burgeoning affluence and openness. Themes of their work and narratives of loss in the artists' work of the 1990s were responding to China's hyper-rapid urbanization, the disappearance of rural village life, traditional values and cultural heritage. Art critics and historians argue some of these post-Deng era artists were purposeful in creating works that were accessible to foreigners and international in their appeal (Pollack 2004). This work was highly sought after by international collectors at the turn of the 21st century.

By the mid to late 1990s, a generation of artists emerged in China who saw art primarily as a lucrative profession. The perception of this commercial side of the artist's lifestyle emerged in broadcast media at the time. The author recalls television advertisements for expensive watches or luxurious real estate properties broadcast in Beijing in the early 1990s. In these television commercials, the actor, a casually dressed man with flowing long hair, held a paint brush and palette while standing in front of an easel. The *nouveau riche* Chinese artist wore an expensive watch on his wrist and was surrounded by an opulent interior on a high floor of a luxury residential tower. This portrayal of a rich aspiring artist suggests the intersection of consumerism, capitalism and the arts had penetrated everyday life in urban China. Ironically, few affluent art collectors in China purchased work by the New Wave or avant-garde artists, and typically bought Chinese antiquities or work by prominent foreign artists.

In broad terms, artists in the 1980s were primarily social activists who sought freedom of expression and operated collectively, particularly for public consumption. The 1985 New Wave sought to remake art theory, create a new aesthetic and transform China's art world. With bouts of government censorship and purges as a result of policy-making, especially the June 4th 1989 Tiananmen Square massacre, creative expression moved away from the public domain by the 1990s. By the late 1990s, China's artists were operating individually, some in self-exile living abroad, and their work was being produced for the international art market. Artists were further "territorialized" in the late 1990s when art districts, *yishuqu*, in adaptively re-used industrial buildings emerged in urban fringes and were connected to academic art institutions. At this point, central government policy and nation-building efforts were directed toward cultural production as real estate developers and art dealers commodified China's artists. This contributed to China's efforts to secure its position as a major world player.

Trends in post-Mao China's built environment

Chinese real estate developers, architects and landscape architects were influenced by westernization after Deng's reforms took hold (Xiaodong 2000; Xue 2005; Campanella 2008). China's "urban fever" aligned with the 1980s Cultural Fever movement and had a major impact on the allied professions working the built environment (Yu & Padua 2007; Friedmann 2005). Between the late 1980s and 2002, China grew from

324 cities to over 650 cities – most with populations in the millions (Padua 2014). At the turn of the 21st century, over 40% of China's nearly 1.4 billion population were living in urban areas (Friedmann 2005).

The Cultural Revolution and ten-year closure of schools left many of China's architecture and landscape architecture programs outdated – lagging far behind western schools in technological knowledge largely standardized by the 1980s internationally. China understood their deficiencies and highly selected students were sent overseas in the early post-Mao period. As design and engineering programs reopened in China's universities, their related Design Institutes were soon under pressure to meet the demands of rapid urbanization in China (Xiaodong 2000; Rowe & Seng 2002; Xue & Ding 2018).

Architects, landscape architects and urban planners trained in China's Mao era were accustomed to utilitarian approaches. When faced with new demands for urban development in the 1980s and 1990s, they attempted to reproduce the various styles that were popular in Europe and North America. Often they fell back on Greco-Roman or neo-classical architecture and the Beaux Arts tradition left over from the Republican period (Xue 2005; Yu & Padua 2007). The post-Mao urban fabric including government and commercial buildings, research and development industrial parks and public open spaces (tree-lined streetscapes, squares and parks) were designed by locally trained designers. Building styles varied and included modern, postmodern and neo-classical, with park designs often within the Chinese Picturesque style. Many of the buildings were poorly executed and construction was often sub-standard (Xiaodong 2000; Xue 2005).

Much of this built work was intended to display the status and importance of local governments or new private enterprises. These ambitions by local politicians were expressed as having the "longest bridge, tallest building, largest convention center" (Xue 2005, p. 6). This gave local design institutes the opportunity to design buildings that moved away from the utilitarian traditions inherited from the Mao era. However, much of this work in cities across China mimicked western styles and were considered superficial (Xue 2005; Xiadong 2000; Yu & Padua 2007). At the same time, many buildings and urban fabric with cultural and historic value were lost. A professor from Southeast University's School of Architecture summarized the situation with a saying popular among architectural preservationists: "there is no construction without destruction in China, *bupo bu li*".

By the turn of the 21st century, several major buildings in China's urban centers were designed by foreign architects who were selected through design competitions. Government business regulations required foreign architects to collaborate with local Chinese design institutes. Many of the design institutes were affiliates of universities or local governments. This provided a mechanism for importation, transfer and absorption of new technologies and ideas (Xiaodong 2003; Xue 2005). For example, local architects were strongly influenced by the innovative work of foreign architects who had created Shanghai's skyline in Pudong (Perry 2004). Like their commercial counterparts in the visual arts, local architects soon began to gain recognition for innovative work in China. Twelve Asian architects were selected to design twelve different buildings for "the Commune", a hotel complex made up of large villas outside Beijing near the Great Wall at Bayading. The architects were subsequently invited to exhibit at the Venice Architecture Biennale in 2002, and the real estate developer, Ms. Zhang Xin, received an award there for her patronage (Xiao 2007).

Chinese designers began questioning the idea that "foreign ideas are always superior" (Xue 2005). After twenty-five years of learning from foreign influences – especially American urbanism – municipal officials, local architects and city planners began to question whether this was short-sighted. Spurred by the loss of historic urban fabric in many large cities, a debate emerged around ways that Chinese identity could be expressed in architecture. This includes a long-standing architectural issue referred to as the "big roof" controversy. In this debate, so-called modern buildings should contain the traditional Chinese big roof with the building's body reflective of the western modern style (Rowe & Seng 2002; Xiaodong 2003). In this architectural rendition, a modern building's roof line would represent *Chineseness* or Chinese identity.

International architects have treated China as an experimental lab to test new ideas, but Chinese architects in the 1980s and 1990s approached the design of their built environment differently. By the turn of the 21st century, these architects typically embraced the idea that China is part of the larger world, but were impassioned about evolving a built form that reflects contemporary China rather than international trends in design alone.

For those Chinese architects who studied or traveled overseas, exposure to international standards of building made them more aware of the limits of their educational system and the need to improve building construction standards. In the 1990s, the Chinese Society of Architects regularly sent delegations to travel internationally and visit universities and private offices abroad with the goal to study best practices. Students were also selected to study abroad through national competitions and mandated to return to teach and practice. By the late 1990s, the government passed regulations for the certification of architects; and a few Chinese architects who were educated overseas established their own private firms and designed significant buildings in China. Some Chinese architects remained abroad, and at one point occupied prominent positions in universities outside China.

The post-Mao era has been equated with westernization in both the arts and architecture. Much of the development of the arts and architecture in the period since the reforms has taken place in the shadow of the highly developed arts of Europe and North America. This has been met with considerable ambivalence. The ambivalence has been made greater by the fact that westernization in China has meant a rush to consumerism and materialism, and it has been accompanied by degradation of the environment and growing inequality. All of this has taken place against the backdrop of the pervasive cultural nationalism that has served to weld Chinese society together for two millennia.

This has produced a contemporary debate over Chinese identity and culture that has become a major influence on the arts and architecture. In certain ways, this debate also recalls issues raised earlier during the May Fourth movement. Chinese artists and designers seek a means of expressing or constructing an identity that is simultaneously rooted in local culture and capable of absorbing and using the most advanced ideas and techniques available in the rest of the world (W. Wang 2006, pers. comm., 21 October).

Post-Mao landscape architecture: foundations of professional practice and education

The discipline of landscape architecture has its own complexity and appears closer in comparison to debates surrounding Chinese literati art, especially garden-making. Like the arts, landscape architecture came to the post-Mao reform period with an established garden tradition and the Chinese Picturesque. In fact, it is widely accepted that China's architectural heritage emerged out of China's classical garden design tradition. This tradition was also entrenched in a tier of educational institutions in China. It survived the Cultural Revolution to emerge intact when universities re-opened after the reforms (X. Sun 2008, pers. comm., 18 June). This context is made more complex by the close relationship between designed landscapes and China's rapid urbanization.

Deng's reforms marked a period of dramatic urban growth in China, especially the rise of secondary cities in the 1990s. It's important to note the reforms occurred in two stages, with 1978–1984 focused on agricultural and land reforms, followed by urban reforms (Wang 2003). Rapid urbanization and the rise of secondary cities signaled growing local government affluence, and fueled a building boom for new urban parks – creating a tremendous demand for new designed landscapes. The interaction of these different forces – sudden growth in demand, an established tier of schools devoted to garden design, growing international awareness and hunger for international ideas, and deeply ingrained cultural nationalism – have helped set the foundation for the profession and practice of landscape architecture in China.

New attitudes toward leisure emerged as part of Deng's reforms and the shifts in economic and political power. The old Communist principle that leisure should serve to promote political harmony and social hygiene was relaxed and spare time became the property of individuals. Under Deng's leadership and throughout China's late 20th century, leisure time could be used as people pleased as long as it did not threaten public order.

1982 was a milestone year in China when the adoption of their revised Constitution took place along with the launch of a new Law on the Protection of Cultural Relics. The latter proved significant to landscape architecture in both education and practice. It defined the protection of cultural relics in a comprehensive yet ambiguous way. For example, sites related to revolutionary history, memorial buildings, sites of ancient culture, ancient tombs, ancient architectural structures, cave temples, stone carvings and the like were designated as sites to be protected for their historical and cultural value at different levels according to their "historical, artistic or scientific value". Hence, post-Mao China's earliest landscape architecture programs took their cues from this law and focused on the preservation of cultural landscapes (X. Sun 2008, pers. comm., 18 June; K. Yu 2007, pers. comm., 8 September).

Around this time, the State Council designated forty-four sites as the first group of China's National Scenic Parks in 1982. However, the loosely defined notion of cultural relics soon became an umbrella for the development of public parks around ancient sites, scenic areas and restoration of traditional gardens. The new designed landscapes built in the 1980s involved pedestrian path improvements in some of the designated forty-four national scenic areas and some new urban parks. These new designs usually depicted the Chinese Picturesque, Beaux Arts or neo-classical genres.

Two examples of 1980s urban parks, Wall City Park in Hefei, the capital city of Anhui Province, and Nie Er Park in Yuxi, an industrial city near Kunming, the capital city of Yunnan Province, were given national awards of design excellence in landscape architecture by the Ministry of Construction (Zhu 1997). The local design institute utilized the Chinese Picturesque genre for Hefei's linear park that marks the 12th century alignment of the city's walls from the Song dynasty. The Yuxi public park commemorates Nie Er, a city native and musician who composed the PRC's national anthem and was considered a local hero. According to China's standard classification of parks, both are considered cultural parks. Yuxi park combines elements of the Chinese Picturesque and axial geometry of the Beaux Arts tradition.

Landscape architecture education in post-Mao China evolved through various topical streams including ornamental horticulture, forestry and urban greening, the classical arts and traditional garden design, as well as engineering and urban planning. As mentioned in an earlier chapter, China's higher institutions of learning were in transition at the turn of the 20th century when attempts were made to modernize or westernize educational practices away from merely passing the imperial civil service examination.

While garden design was practiced across China's dynastic period, it wasn't until the first decades of the 20th century when horticulture and garden design were taught in a few schools in China by Chinese educators who were schooled in France and Japan (Lin 2005). The architecture school at Southeast University (formerly known as National Central University) in Nanjing offered courtyard design by the late 1920s and courses on landscape gardening or garden design were offered at an agriculture school in Suzhou. Later in the Mao era, so-called scenic garden design was offered at Shanghai's Fudan University circa 1949 and established in Beijing Agricultural University circa 1951 (X. Sun 2008, pers. comm., 18 June). Sun (2008 interview) noted as Mao's "Greening the Motherland" was launched, scenic garden design or landscape architecture was moved from Beijing Agricultural University to the new Department of Urban and Residential Greening at Beijing Forestry University circa 1956. As mentioned in a previous chapter, Soviet-style urban parks were built during the Mao era. However, with some efforts focused on voluntary tree plantings by the masses and agricultural experiments, along with Mao's anti-aesthetic, utilitarian-based perspectives and period of economic scarcity, landscape architecture didn't flourish then as a professional discipline.

China's designation of the forty-four national scenic areas in 1982 appears to have influenced the expansion of landscape architecture education amidst the larger re-shuffling of programs throughout the older and established universities. For example, at the inland capital city of Hubei Province along the Yangtze River, Wuhan University's Landscape Gardening program moved to the engineering program in the School of Urban Construction at Wuhan University of Science and Technology, formerly Hubei Technical Institute (D. Li 2010, pers. comm., 22 November). Later, Tongji University in Shanghai also re-classified their landscape gardening program as part of engineering in 1987 (B. Liu, 2009 pers. comm., 18 May). In fact, so-called "landscape gardening" programs were officially designated as engineering by the Ministry of Education in 1987 (Lin 2005; Yu 2007). Ten years later, the curriculum of landscape gardening was inadvertently eliminated by China's Ministry of Education and subsumed into urban planning programs (D. Li 2010, pers. comm., 22 November). Educators interviewed believe that moving the landscape gardening programs into Urban Planning Departments terminated the conventional development of landscape education from horticulture or forestry. This allowed for some change and innovation where studies shifted from the unit scale of classical gardens and scenic design to the larger city scale.

Landscape architecture education since then has taken many forms of undergraduate studies and is typically offered in schools or departments of agriculture, forestry, horticulture, architecture and engineering. This is not unlike the tradition of landscape architecture education in North American and western European universities. The following offers some highlights of changes in landscape architecture education since it was moved to urban planning programs in China's universities in the late 1990s.

Yu Kongjian, now a world-known award-winning landscape architect and founder of Turenscape, China's first private professional practice, was one of the earlier Chinese nationals who studied landscape architecture overseas at Harvard University. Yu established the Center for Landscape Architecture and Planning at Peking University's Department of Geography in 1997. Eventually, Yu established Peking University's Graduate School of Landscape Architecture, offering post-graduate studies in early 2003. Later that year, Tsinghua University's School of Architecture established their Department of Landscape Architecture with Laurie Olin, an internationally known American landscape architect, serving as its founding head. Peking University's Master of Landscape Architecture professional degree was certified by China's Ministry of Education in June 2005 and set the national precedent for graduate-level or post-graduate professional landscape architecture programs in China.

The concept of landscape architecture as a modern profession in China was highly contested and controversial at the turn of the 21st century, and appears to continue today. For some, it follows an evolutionary trajectory originating with traditional garden design; and for others, like Yu Kongjian, the professional practice of landscape architecture was not unlike China's 1985 New Wave or avant-garde artists. Their desire, as mentioned earlier, was to modernize and remake art theory, as well as create a new aesthetic for the arts in China, one that is innovative and breaks with traditional literati art. Yu, a self-styled avant-gardist, seeks to demonstrate to Chinese society a new aesthetic that represents the intersection of cultural and natural resources, especially as a response to China's irreparable environmental destruction due to decades of unregulated growth.

Despite the debate on whether landscape architecture practice evolved from traditional garden design or completely breaks with this past tradition, the Ministry for Labor and Human Resources officially announced the establishment of landscape architecture as a profession in December 2004. It was to be considered an independent profession, parallel to architecture and planning. Regulations for the certification of architects and urban planners were instituted in 1995 and 1999, respectively. However, the creation of regulations for the certification of landscape architects has been slow in the making. Since China's rapid urbanization, new programs in landscape architecture have emerged in a variety of schools around China. This was particularly compounded by a central government plan for rural to urban in-migration – forecasting 65% of the nation's population in cities by 2020. The tremendous diversity in the range and type of landscape architecture programs were represented at a 2007 conference for Chinese landscape educators hosted by Peking University. There, the author met landscape educators who were based in diverse fields: fine art programs in government art academies, private art institutes, engineering, industrial design, architecture, forestry, agriculture, tourism and natural sciences. No form of accreditation has yet been put in place for these landscape architecture programs.

Landscape architecture "outside" post-Mao China

By the time China opened to the world in the late 1970s, the profession of landscape architecture was operating in various modes of practice, especially in the United States. An overview of design trends "outside" China in the western landscape architecture praxis provides a benchmark for China's hybrid modernity. It informs ways international design trends may have influenced post-Mao China's designed land-scapes, especially public parks in urban contexts.

In the USA, the profession of landscape architecture was established with Frederic Law Olmsted's 19th century New York Central Park and blossomed in the post-World War II period with the rise of the so-called American corporate firms, like EDAW, Sasaki Associates, WRT, and the SWA Group before China opened to the world. Lawrence Halprin (1916–2009), award-winning California-based landscape architect, had a staff of nearly 100 employees in the 1970s. Peter Walker, a founding partner of Sasaki, Walker and Associates (SWA), a firm created circa 1957 departed to establish Peter Walker Partners in 1983. These landscape architects and corporate firms emerged during a period of American economic growth after World War II and contributed to the professional development of landscape architecture beyond the scope of residential garden design. The projects they worked on involved planning and development, natural resources conservation and management, as well as the design of medium and large-scale projects for clients in both the public and private sectors.

Projects included brownfields reclamation and conversion of former industrial sites to public parks, the development of open space systems, related master plans and regional open space systems of parks, recreational and leisure spaces in American cities and counties, as well as the range of new school campuses. Additionally, northern California's "silicon valley" and computer technology sector had just emerged within proximity of Stanford University in the late 1970s when the area was transformed into a suburban landscape of so-called "R + D (research and development) light industrial parks". Some of these projects were comparable in size to small to medium-sized college campuses. Numerous residential suburban developments in this area and throughout the United States were also built during this time. To meet the demands of the growing middle class and corporate executive class, vacation and leisure resorts were developed in Hawaii, Mexico, and the Caribbean islands, for example, as well as ski resorts in North American mountain locations. By the 1980s, corporate planning, design and engineering firms were engaged in the development of new towns in the Middle East, and related to that region's oil boom.

The Disney Corporation expanded its original southern California theme park circa 1950s to Orlando, Florida; Paris, France; and Tokyo, Japan. This period also witnessed the demand for second homes by corporate executives, and vacation condominiums for the American middle and upper-middle class, as well as golf course-oriented residential communities at Hilton Head Island, South Carolina (circa 1960s), Florida and other locations in the USA. During the early 1980s and 1990s, American landscape architects were often leading large multi-disciplinary teams for major mixed-use commercial, retail and suburban or resort-oriented residential development projects, and other complex projects like the reclamation of brownfields.

In the late 1970s, landscape architect Martha Schwartz transformed the seemingly monotonous design signature of corporate firms with her breakthrough, Bagel Garden, a site-specific project inspired by Dada, a conceptual genre in the fine arts. By the mid-1980s, a new generation of landscape architects interested in creating a more artful design aesthetic came to light. Some emerged out of the corporate realm or large consulting firms, like George Hargreaves who worked at the SWA Group or Michael Van Valkenburgh who worked at Carr, Lynch Associates, Inc. Other firms like Pamela Burton and Company opened their own design office by choice.

The international world of landscape architecture shifted with Bernard Tschumi's winning entry for the 1982–1983 international design competition for *Parc de la Villette*, at the former site of Parisian slaughterhouses and one of President Francois Mitterand's *Grands Projets*. Typical of projects and design competitions in France, a philosopher is engaged. In keeping with this practice, Bernard Tschumi, Swiss architect, sought out Jacques Derrida, a postmodern philosopher who developed the notion of deconstructivism. Postmodernism was a trend in architecture throughout the west at the time and it appeared for the first-time in Tschumi's park design. This postmodern park was completed by 1987 and created controversy among conservative landscape architects. Design theorists and critics in the discipline of landscape architecture were quick in their analysis of Tschumi's proposal for *folies* in a system of points, lines and planes (Meyer 1991; Baljon 1992). In some ways, Tschumi's proposal expanded on the 1980s new wave of landscape architects who were interested in meaning and visual quality. He shifted the notion of meaning by formalizing a design approach based on deconstructivist philosophy.

Peter Connolly, Australian landscape architect and educator, currently at New Zealand's Victoria University in Wellington, coined the term landscape urbanism in the early mid-1990s (interview L. Corkery, 10 May 2010). Subsequently, Charles Waldheim and others reinforced this theory to re-orient landscape architecture to the city. Instead of buildings as the form-drivers of the city's fabric, landscape urbanism postulates open spaces or designed landscapes as the physical determinants of a city. Over time, landscape urbanism and landscape architecture practice morphed into the realm of green infrastructure, stormwater management and focus on the performative dimensions of designed landscapes within cities and suburban contexts – diverging away from the importance of aesthetics. Eventually, landscape urbanism evolved or branched off into ecological urbanism, a type of natural systems or ecological approach to designing cities.

By the turn of the 21st century, James Corner and his firm Field Operations expanded the discipline further with their winning entry for the 2001 international design competition for Fresh Kills Park, a long range master plan to transform New York city's 2200 acre landfill site into a series of environmentally safe public parks. Fresh Kills was an active toxic landfill site that contained over 150 million tons of New York city's trash until the city decided to close and de-activate it in 2001. Corner's master plan proposed a series of landscape interventions that over time would restore the site's natural landscape (meadows, woodlands and marsh) and create places for recreation (soccer fields, ballfields, children's playgrounds, hardball courts) for the adjacent neighborhoods, a greenway, as well as a path system for hikers, walkers and bicyclists. Corner's project also created new habitats and the area emerged as a field laboratory for natural scientists studying the variety of birds, turtles and other animals.

Essentially, when China opened to the world in the late 1970s, and more fully by the mid-1980s, the international discipline of landscape architecture had an established professional presence with lively debates about innovation and the avant-garde. The United States' post-World War II period of growth caused the rise of corporate landscape architecture firms as China traversed through Mao's chaotic period of extremes, contradictions, violence and destruction, and global isolation. As China fully opened to the world by the mid-1980s, deconstructivism and Tschumi's postmodern design for *Parc de la Villete* created controversy in the established academy of landscape architecture. In the late 1970s, Martha Schwartz's Bagel Garden had also upset the established American community of landscape architects. By the turn of the 21st century, landscape urbanism and the performative qualities of designed landscapes (green infrastructure, best management practices for stormwater management) in cities dominated the landscape architecture discourse in the west.

When China opened to the world, landscape architecture as a modern and contemporary profession was in its embryonic stages there, and by the turn of the 21st century had emerged as an international living laboratory for urban experimentation. Then, China's new 1980s designed landscapes were often representative of the Chinese Picturesque, especially as sites of cultural significance were improved. Some projects drew from the neo-classical and Beaux Arts, a style in vogue during the earlier 20th century. Innovation in Chinese park design largely occurred by the 1990s as a result of hybrid modernization, related hyper-rapid urbanization and consequential environmental degradation. Various central government policies were aimed at nation-building, and at the same time created competition among city mayors with the rise of secondary cities. Political ambitions among local leaders in China's secondary cities drove innovation in park design.

The post-Mao construction of public parks was instrumental to legacy-building for local officials. These parks were created as city landmarks or icons and critical locations for foreign direct investment. These new purpose-built public parks also demonstrated a measure of local officials' success to central government, as part of China's government assessment process that advanced local officials up the political ladder. Government officials in four secondary cities enabled the innovation of four distinctive designs of purpose-built parks in the last decade of 20th century China. These four projects: Living Water Park in Sichuan Province; Zhongshan Shipyard Park in Guangdong Province; Jinji Lake in Jiangsu Province; and West Lake's Southern Scenic Area in Zhejiang Province represent China's late 20th century hybrid modernization. These projects serve as the vanguard for the development of China's contemporary discipline of landscape architecture and the schema for hybrid modern design. Soon after each park was completed, China's central government placed these four purpose-built parks on a mandatory list of exemplars for central and local government officials to visit. The next chapter synthesizes hybrid modernity through the case study analysis of these parks and discursive narrative for late 20th century design innovation in China.

References

Andrews, J. (1994) *Painters and Politics in the People's Republic of China: 1949-1979*. Berkeley, CA, University of California Press.

Apter, D. (1993) Yan'an and the Narrative Reconstruction of Reality, *Daedalus*, 122(2), 207–232.

Baljon, L. (1992) *Designing Parks*. Amsterdam, Netherlands, Architectura and Natura Press.

Barmé, G. (1999) *In the Red: On Contemporary Chinese Culture*. New York, NY, Columbia University.

Berghuis, T. (2012) Experimental Art, Performance and 'Publicness': Repositioning the Critical Mass of Contemporary Chinese Art, *Journal of Visual Art Practice*, 11(2+3), 135–155.

Calhoun, C. (1989) Tiananmen, Television and the Public Sphere: Internationalization of Culture and Beijing Spring of 1989, *Public Culture*, 2(1), 54–71.

Campanella, T. (2008) *The Concrete Dragon*. New York, NY, Princeton Architectural Press.

Carrico, K. (2017) Eliminating Spiritual Pollution: A Genealogy of Closed Political Thought in China's Era of Opening, *The China Journal*, 78, 100–119.

Cheek, T. (1984) The Fading of Wild Lilies: Wang Shiwei and Mao Zedong's Yan'an Talks in the First CPC Rectification Movement, *The Australian Journal of Chinese Affairs*, 11, 25–58.

Davis, D. S. (1995) Introduction: Urban China. In: *Urban Spaces in Contemporary China: The Potential for Autonomy and Community in Post-Mao China*, Davis, D. S., Kraus, R., Naughton, B., & Perry, E. (eds.). Washington, DC, Woodrow Wilson Center Press.

Devin, D. (1997) *Mao Zedong*. Stroud, Gloucestershire, UK, History Press.

Dillon, M. (1998) *China: A Cultural and Historical Dictionary*. London, UK, Curzon Press.

Dirlik, A. (2002) Modernity as History: Post-Revolutionary China, Globalization and the Question of Modernity, *Social History*, 27(1), 16–39.

Flath, J. (2004) It's a Wonderful Life: Nianhua and Yuefenpai at the Dawn of the People's Republic, *Modern Chinese Literature and Culture*, 16(2), 123–159.

Friedmann, J. (2005) *China's Urban Transition*. Minneapolis, MN, University of Minnesota Press.

Gao, M. (1999) From Elite to Small Man: The Many Faces of a Transitional Avant-Garde in Mainland China. In: *Inside/out: New Chinese Art*, Gao, M. (ed.). San Francisco, CA, San Francisco Museum of Modern Art.

Gao, M. (2011) *Total Modernity and the Avant-Garde in Twentieth-Century Chinese Art*. Cambridge, MA, The MIT Press.

Goldman, M. (1981) *China's Intellectuals: Advise and Dissent*. Cambridge, MA, Harvard University Press.

Hung, C. T. (2011) *Mao's New World: Political Culture in the Early People's Republic*. Ithaca, NY, Cornell University Press.

Li, X. (1993) Major Trends in the Development of Contemporary Chinese Art. In: exhibition catalog, *China's New Art, Post 1989, with a Retrospective from 1979–89*, 31 Jan–25 Feb 1993, Hong Kong, Hanart Z Gallery.

Liao, W. (1993) Unrepentent Prodigal Sons: The Temper of Contemporary Chinese Art, In: exhibition catalog *China's New Art, Post 1989, with a Retrospective from 1979–89*, 31 Jan–25 Feb 1993, Hong Kong, Hanart Z Gallery .

Lin, G. S. (2005) Review and Prospect: A Study of Landscape Architecture Education in China, *China Landscape Architecture Journal*, 9 and 10, 1–9, 73–79.

Liu, X., Sun, S. H. & O'Connor (2013) Art Villages in Metropolitan Beijing: A Study of the Location Dynamics, *Habitat International*, 40, 176–183.

McDougall, B. S. (1980) Mao Zedong's "Talks at the Yan'an Conference on Literature and Art" A Translation of the 1943 Text with Commentary". In: *Michigan Papers in Chinese Studies*, 39. Ann Arbor, MI, Center for Chinese Studies, The University of Michigan.

Meserve, W. J. & Meserve, R. I. (1992) Revolutionary Realism: China's Path to the Future, *Journal of South Asian Literature*, 27(2), 29–39.

Meyer, E. (1991) The Public Park as Avant-Garde, *Landscape Journal*, 10(1), 16–26.

Padua, M. (2014) China: New Cultures and Changing Urban Cultures. In: *New Cultural Landscapes*, Roe, M. & Taylor, K. (eds.). London, UK, Routledge.

Pollack, B. (2004) Beyond Stereotypes, *ARTNews* 103(2), 98–103.

Perry, E. (2004) Shanghai's Political Skyline. In: *Shanghai Architecture and Urbanism for Modern China*, Rowe, P. & Seng, K. (eds.). Munich, Germany, Prestel.

Rowe, P. & Seng, K. (2002) *Architectural Encounters with Essence and Form in Modern China*. Cambridge, MA, MIT Press.

Selden, M. (1995) Yan'an Communism Reconsidered, *Modern China*, 21(1), 8–44

Sheng, H. (1990) Big Character Posters in China: A Historical Survey, *Journal of Chinese Law*, 4(2), 234–256.

Spence, J. D. (1991) *The Search for Modern China*. New York, NY, W. W. Norton & Company.

Sullivan, L. (1988) Assault on the Reforms: Conservative Criticism of Political and Economic Liberalization in China, 1985–86, *The China Quarterly*, 114, 198–222.

Sullivan, M. (1999) Art in China Since 1949, *The China Quarterly*, 159, Special Issue: The Peoples Republic of China after 50Years, 712–722.

Vaughan, J. C. (1973) *Soviet Socialist Realism*. New York, NY, Palgrave Macmillan.

Wang, H. (2001) Chinese Contemporary Thought and the Question of Modernity. In: *Whither China? Intellectual Politics in Contemporary China*, Zhang, X. (ed.). Durham, NC, Duke University Press.

Wang, H. (2003) *China's New Order: Society, Politics, and Economy in Transition*. Cambridge, MA, Harvard University Press.

Xiao, C. (2007) The Dynamic Duos Daring to Dream, *China Daily*, 9 February.

Xiaodong, L. (2000) Celebration of Superficiality: Chinese Architecture Since 1979, *Journal of Architecture*, 5(3), 391–409.

Xiaodong, L. (2003) Implications of Architectural Education in China in contemporary Chinese architecture, *Journal of Architecture*, 8(2), 303–320.

Xue, C. Q. L. (2005) *Building A Revolution: Chinese Architecture Since 1980*. Hong Kong, Hong Kong University Press.

Xue, C. Q. L. & Ding, G. (2018) *A History of Design Institutes in China: From Mao to Market*. Abingdon, UK and New York, NY, Routledge.

Yu, K. (2007) The evolution of landscape architecture in China. In Lffage, A. (ed.) *Proceedings of Between Architecture and Landscape Education*. Paris, France, Ecole Nationale Supérieure d'Architecture de Paris La Villette, 22–23 Nov.

Yu, K. & Padua, M. (2007) China's Cosmetic Cities: Urban Fever and Superficiality, *Landscape Research*, 32(2), 225–249.

Zhang, X. (1997) *Chinese Modernism in the Era of Reforms*. Durham, NC, Duke University Press.

Zhong, Y. & Cui, Y. (2015) Study on Higher Education Fund Guarantee System of the Communist Party of China in Yan'an Period, *Higher Education of Social Science*, 8(2), 34–39.

Zhu, G. H. (1997) *Landscape Architecture Designs of Excellence: 1986–89 Vol 2*. Tianjin, China, Tianjin University Press.

5 Revealing late 20th century hybrid modernity

Four purpose-built parks

The concept of hybrid modernity is directly related to late 20th century hybrid modernization, a socio-cultural process that emerged when Deng Xioaping's reforms opened China to the world. Major critical shifts contributed to this process in the aftermath of Mao's period of extremes, contradictions, and unpredictability. China opened to the world after decades of isolation and the social rupture of Mao's Cultural Revolution when urban intellectuals were publicly humiliated, and all schools were closed for ten years and urban dwellers sent to the countryside to learn from the peasantry. China's largely rural population also suffered from severe famine and poverty as a result of Mao's failed policies. Deng's reforms and "four modernizations" called for decentralizing central government and shifting fiscal autonomy to local governments. This new political context enabled the rise of secondary cities where local officials became entrepreneurial and their cities became vessels for economic development, foreign investment and places for tremendous economic growth and China's "urban experiment." Chinese language translations of a variety of published materials and media from postmodern western nations flooded China and contributed to their overall socio-cultural transformation. At the same time, China's intelligentsia and society, especially university students, grappled with cultural identity as the Communist Party of China (CPC) and nation-building efforts transformed society. This unprecedented government restructuring, collision of "local" cultural identity and international or "global" influences, and related hyper-rapid urbanization informed the process of hybrid modernization, socio-cultural dimensions of post-Mao late 20th century China. Hybrid modernity, the representative design genre, is revealed through the case study analysis of four purpose-built public parks located in China's secondary cities. As Francis (2001) notes, "case studies can be instrumental in developing new theories related to landscape architecture". This chapter's case study analysis serves as the empirical evidence for the theoretical construction of post-Mao China's hybrid modernity.

It's important to reiterate China's secondary cities were the places where China's hybrid modernization took hold. With the shift to local government autonomy and changes to land tenure and property rights, local officials became both entrepreneurial and competitive. To advance politically, local officials were evaluated based on their achievements. Greening or afforestation and new purpose-built public parks were important achievement indicators for the central government. Local officials understood ways new public parks could increase real estate property values, make their cities more attractive for foreign investment, and demonstrate environmental sustainability and community well-being. The construction of new public parks was inexpensive, especially compared to new buildings, and readily executed in secondary cities. In contrast, China's primary cities, Beijing, Shanghai and Guangzhou, were already places with an international reputation with their own park heritage, and their political leaders had different needs and metrics for evaluating success. China's secondary cities were places where local (municipal, provincial and CPC) leaders had the flexibility to innovate during the tremendous late 20th century urban experiment. New public parks and urban greening contributed to local cultural identity, nation-building, community development and well-being.

As stated earlier, the case study analysis draws from field research, archival review of original design drawings, and design vocabulary and language from both international design trends in landscape architecture and the Chinese Picturesque garden tradition. Content analysis of interview responses from Chinese government officials and experts including design educators and professionals working in China was also considered. The four case study public parks were initiated in the late 1990s with varied project gestation and construction times. Living Water Park in Chengdu, Sichuan Province was completed in 1998 and the others in the early years of the 21st century. Deng's initial reforms focused on changes to the land tenure system and rural areas in the 1980s. By the 1990s, unregulated urbanization had accelerated tremendously, especially the rapid growth of China's secondary cities as entrepreneurial places where local leaders were reimagining local identity and branding their cities for foreign investment. This chapter's discursive narrative examines the four case study purpose-built parks individually and chronologically. It takes into account physical, historical and cultural contexts; background and inception of the park's design with perspectives from respective clients; designers' intentions; project realization; and experiential qualities. Concepts of nation-building and local identity discussed in previous chapters are also considered in the discursive narrative for hybrid modernity. The chapter closes with a synthetic discussion of the four purpose-built hybrid modern public parks:

- Living Water Park, *Huoshui Gongyuan*, Chengdu, Sichuan Province (1995–1998)
- Zhongshan Shipyard Park, *Qijiang Gongyuan*, Zhongshan, Guangdong Province (1998–2001)
- Jinji Lake Landscape Master Plan, *Jinji* (Golden Rooster) *Hu Jingguan Zongti Guihua*, Suzhou, Jiangsu Province (1998–2002)
- West Lake Southern Scenic Area Master Plan, *Xihu Huanhu Nanxian Jingqu Zongti Guihua*, Hangzhou, Zhejiang Province (1999–2002)

Case study one: Living Water Park, *Huoshui Gongyuan*, Chengdu, Sichuan Province

> "my inspiration lies in the sacred water sites, those places that people who honored the power and the need for living water considered core to their survival".
>
> Damon (2008, pers. comm., 25 July)

Living Water Park (LWP) is located in the Chengdu, capital city of Sichuan Province in southwestern China (Figure 5.1). LWP grew out of a larger municipal capital works modernization project known as the Funan 10-year Comprehensive Revitalization Plan. This larger project involved cleaning the polluted Fu and Nan rivers' corridor and mitigation measures to eliminate the pollution sources. The project's intentions sought to modernize Chengdu's public infrastructure system and improve the city's river water quality and arrest environmental degradation.

Chengdu is referred to as a secondary city in this study – an important local or regional center, not a top-tier primary city or internationally known city like Beijing,

Figure 5.1 Context and location map for Living Water Park, graphic by X. Liu

Shanghai or Guangzhou (formerly Canton). Chengdu and the Sichuan Province are geographically located in China's western region and are situated in an agricultural area known as the Sichuan plain. Sichuan Province is south and east of the Himalayas with Chengdu situated within the Min River, *Min Jiang*, watershed area that stretches from western Sichuan into Tibet. The Min River is a major tributary of the Yangtze River, *Chang Jiang*, the longest river in China that flows into the delta region and East China Sea near Shanghai. The Min River was the area's lifeblood for thousands of years and was sacred to the local population (Sage 1992). More than 2000 years ago, its waters served as the basis for the Dujiangyan Irrigation System, China's first man-made water-based engineering system designated a UNESCO World Heritage site circa 2000. The Min River is also the natural source of the Fu and Nan rivers, which in turn historically supplied the moat around Chengdu's ancient city walls and defined the boundaries of the former historic walled city. For purposes of brevity, these two rivers will be referred to as the "Funan" for the balance of the chapter.

LWP was built on a central Chengdu site (5.9 acres/2.4 hectares) along the reconstructed Funan riverbanks and intended as a memorial to the city's flagship modernization project. The Funan Rivers Revitalization Bureau, a local government administrative unit, was created to oversee the city's larger modernization project. The scope of this modernization project was complex and included dredging and reconstructing the riverbanks; demolishing existing buildings (sub-standard housing and factories) and eradicating all pollution sources along the river corridor; installing new flood control management and a new network of public infrastructure with new roads and bridges; and new housing and urban greening. As noted in an earlier chapter, the term "urban greening" in China generally refers to afforestation and greenbelts, tree-lined roads, public parks and squares. In the process of the city's modernization project, the city added several acres of public green space and planted over 20,000 trees (Padua 2004b).

Chengdu context

Before delving further into LWP, a brief overview of Chengdu's history sheds light on the project's inception. Chengdu's history played a significant role in the evolution of China's cultural and political development. It was the capital of Shu, circa 220–280 CE, one of the Three Kingdoms in the period of disunity after the Han dynasty. It was the birthplace for the Shu silk brocade, a major commodity that sustained the economy of the city. Chengdu is purportedly the birthplace for China's traditional tea culture (Sage 1992). Woodblock printing and engraving typography were first used in the region (Clunas 1997). Later, it was a commercial hub along the ancient "Southern Silk Road" and linked China to the Middle East. The birthplace of Daoism is located in nearby Qingcheng Mountain, *Qingcheng Shan*, and as a sacred Daoist pilgrimage site contributed to Chengdu's regional importance.

During the Tang-Song era (circa 618–1279 CE), Chengdu was known as an educational and cultural hub for literature and the arts. Local craftwork, arts and literature flourished in Chengdu over a period of fourteen centuries through the late Ming dynasty (Mote 1970). Local legend claims that the luster and sheen of the Shu silk brocade, a major commodity, was a result of being washed in the pristine waters of the Funan (Sage 1992). This contributed to the river's reputation as the city's symbol. Funan was also known historically as the "Brocade River, *Jinjiang*" (Qin 2015). The local community still refers to the river by its historic name, *Jinjiang*. A combination of mercantilism, the silk trade, and Daoism sustained Chengdu's regional importance until the Opium War when it declined. Additionally, its late Qing dynasty decline was a result of a political and geographic shift within Sichuan Province when the capital city moved from Chengdu to nearby Chongqing, circa 1800s (Skinner 1977).

Chengdu was later re-established as the capital city of Sichuan Province during the early 20th century modern Republic of China (ROC) period. Civil war and the Japanese occupation soon made Chengdu and nearby Chongqing places of refuge for people who fled Nanjing, ROC's capital. This mass relocation of people to Chengdu – including government officials, members of the military, politicians, merchants, artists and poets – caused Chengdu to re-emerge as a significant regional hub and helped sustain its commercial and cultural vitality through this tumultuous period (Mote 1970). Like most cities in the Mao era, Chengdu declined both culturally and economically.

However, the physical environment of Chengdu remained relatively unspoiled through the 1960s; local residents still fished and swam safely in the Funan (Padua 2004b).

By the late 1970s, people who had been sent to the countryside during the Cultural Revolution returned to Chengdu in Deng's New Era. Local government provided assistance for this relatively small jobless population by constructing factories and simple low-rise housing along the Funan. Unregulated discharge of effluent from this housing and industrial pollution from the factories eroded the quality of the Funan as Chengdu experienced rapid growth in industry and urbanization following Deng's reforms. Like other locations in China, unregulated growth occurred without environmental control. Sub-standard housing expanded along the river corridor and central Chengdu became a magnet for further urban in-migration. Over 100,000 residents lived in the sub-standard housing along the riverfront along with over 1000 unregulated businesses. The combination of sewage from the sub-standard housing and pollutants from factories degraded the Funan significantly and it was known locally as "*Fulan jiang*", – the Rotten River (Padua 2004b).

The catalyst for the massive effort to clean the Funan river corridor originated with a letter from students at the Long Jiang Lu elementary school who petitioned the mayor in 1985. The students, under the guidance of their natural sciences teacher, pleaded for action to clean the polluted river and stop environmental degradation. The former mayor, Hu Maozhu, was deeply moved and challenged by the students' environmental activism expressed in their letter; and his first step was to build a coalition of Chengdu stakeholders who shared the students' concern for Chengdu's future.

In-depth discussions among local government officials, members of the community and representatives of the central government led to a comprehensive set of technical reports and feasibility studies to evaluate the river corridor's pollution and environmental impairment. The results of the studies confirmed high levels of pollution caused by factories and sub-standard housing along the river corridor, and the city's infrastructure had not kept pace with the area's expansion since Deng's reforms. The city's stakeholders also understood the benefits of linking the river's clean-up with new infrastructure as a means of modernizing the city. Furthermore, they realized a comprehensive approach was needed to deal with water quality, the modernization of infrastructure, compatible land uses and environmental control, and the inter-relationships with social and economic development, public health and community well-being. Their vision combined improving the area's natural ecology, reducing water pollution in the upper reaches of the Yangtze River and eradicating poverty to renew the inner city. They sought to promote social and economic development, protect and renovate historic Chengdu, and especially reclaim the river as a symbol of the city's identity.

The modernization project commenced in the early 1990s with the CPC party official, Zhang Zhi Hai, as the project's champion and designated director of the Funan Rivers Revitalization Bureau, an administrative unit created to oversee implementation. Initial work involved the demolition of riverfront factories and sub-standard housing; dredging the river; and re-shaping the river channel for flood control and environmental remediation. It also included the relocation of 100,000 residents from sub-standard riverfront housing; construction of a new consolidated network of major public utilities along the reconstructed river banks; and urban greening. This latter greening component involved the creation of thirteen new public parks called the Green Necklace with one designated park commemorating the city's modernization and river clean-up project (Padua 2004b).

Project inception

As noted in an earlier chapter, American environmental activist and artist Betsy Damon organized and curated international and Chinese artists to engage in two informal public art events on the theme "living water" in Chengdu circa 29 July through 13 August 1995 and in Lhasa, Tibet Autonomous Region circa 18 August through 3 September 1996. The two "unofficial" events involved performance art and temporary public installations, and were not formally approved by the local government. As the curator, Damon's inspiration was based on her 1993 visit sponsored by the Jerome Foundation when she studied local mythology associated with the region's water (Padua 2004b). Her goal for these art events was to raise environmental consciousness about the Funan and Lhasa Rivers and educate local society about water pollution. These events were similar in spirit to the 1980s social activism represented by artists who collectively participated in "Happenings" or performance art in public places throughout China.

In Chengdu, over twenty artists (international and Chinese) participated in and around public areas along the Funan river corridor. As unofficial events, neither Damon nor any of the participating artists sought permits from local authorities. The events were broadcast on local television and publicized widely in print media. Zhang attended the Chengdu events and was emotionally moved by a group performance of artists washing long pieces of white silk in the Funan (Padua 2004b). The longer the artists held the silk in the river the dirtier the silk became. Zhang understood the historical and cultural reference to the river's pristine waters for the Shu silk brocade luster, especially as the ancient capital city's icon during the Three Kingdoms period.

Zhang arranged to meet Damon and informed her about the city's modernization project. After discussions on the city's vision, Zhang invited Damon to design a park that captured the spirit of the art event he witnessed. Damon accepted his invitation with the condition that she received artistic freedom to base the park's design around her life's work on "living water". Zhang agreed to her terms; and Damon later returned to Chengdu in 1996 accompanied by Margie Ruddick, a Philadelphia-based landscape architect who Damon selected after interviewing other landscape architects (M. Ruddick 2006, pers. comm., 12 June).

During their four week collaboration and meetings with city officials, Damon and Ruddick envisioned the new green public space as a place that would bear testimony to the importance of water as the foundation of life and the city's collaborative efforts to clean the pollution along the Funan river corridor and modernize the city. In this context, the park's design intended to educate the local citizenry about the river's original natural environment, how it was damaged and ways it could be improved. Damon and Ruddick refined the city's vision into more specific design objectives: reclaim the memory of the river's symbolism; utilize the water cleansing process as the park's primary form-driver; educate the community about environmental responsibility and local ecology; and celebrate the city's modernization project. Both Damon and Ruddick developed a multi-layered design that combined art, science, culture, and environmental education in a way that was accessible to the community. Their goal was to create a park that would live in the city's memory for future generations while simultaneously creating a place for environmental education (B. Damon 2006, pers. comm., 12 September).

Damon's team included Jon Otto, a physicist and Damon's son, who served as the team's English-speaking translator and also provided technical support, and Huang Shi Da, an ecologist and wetlands expert. After four weeks of intensive design work, the team received the city's approval. During the implementation or construction stages in

China, foreign designers were rarely engaged in this phase of the project at that time. Ruddick returned to the US and Damon remained in Chengdu to work with local sculptor Deng Le on the park's flowforms, water drop fountain, and nautilus sculpture.

LWP has achieved significant national and international recognition since its completion in 1998. It was also listed on China's national tour registry as an exemplar ecological park for government officials to visit and considered the first of its kind in China (X. Sun 2008, pers. comm., 18 June). LWP is considered the first urban park in the world that demonstrated biological and natural processes to treat polluted water (Amidon 2001). However, similar natural processes to purify water from a waste water reclamation plant were deployed in a suburban area south of Seattle, Washington, known as Waterworks Gardens, and completed in 1996. Conceptualized by environmental artist Lorna Jordan, this award-winning eight-acre project was designed as an earth-water sculpture and has five garden rooms where water is funneled, captured and filtered before released into a natural creek. Both, arguably, are early exemplars of ecological urbanism, an evolution of the tenets of landscape urbanism.

The sole sourcing of the LWP design commission to Damon was highly unusual. The conventional procedure for selecting foreign consultants in China during that period was through a selective competitive bid process with invitations made to selected designers by local officials. Damon, in effect, was sole sourced and invited directly by a local administrator and given the freedom to innovate the park's design on approximately seven acres, the largest of Chengdu's thirteen parks in their Green Necklace. Purpose-built parks in Chengdu typically followed local design standards, for example, physically inaccessible lawn areas with security fencing; quiet zones with sitting areas; active zones for exercising and socializing; a formal entry or park gate; designated children's play area; dedicated area with figurative statuary, usually a memorial to a local hero; a path system for both pedestrians and park maintenance vehicles; a nursery or growing area for plants with a utility building; cultural exhibition building; and large assembly area for speeches during national holidays. Given Zhang's commitment to Damon and his authority as the director of the city's Funan revitalization effort, Damon's team was given creative freedom to deviate from these standards.

Design realization

As noted earlier, case study analysis of each of the book's four parks factored in an understanding of the client's and designers' perspectives and intentions. This information was garnered through interviews (in person and via telephone and e-mail correspondence). It also drew from archival (review of technical drawings, reports and design documents) and field research involving site visits.

In the case of LWP, Ruddick indicated her experience was transformative. She immersed herself and learned about local culture and river folklore, especially through her reading of The Book of the Way, *Dao De Jing*. The foreign culture and surroundings enabled Ruddick to focus and effectively collaborate with Damon in a meaningful way. Recognizing it was Damon's first experience to work on a purpose-built park, Ruddick took it upon herself to educate Damon about the landscape architect's approach to natural water cleansing processes. Ruddick described the collaborative experience as part science and professional spontaneity, and reflected on her experience in China: "the spirit of her work not necessarily as wise, but as 'trace-less', where principles of invisibility and collaboration as the product, can be a great form of practice, both in landscape architecture and in life" (10 March 2010, e-mail correspondence).

Acting on their client's goals to educate the community about the river's environment and the city's pollution remediation efforts, Damon's team envisioned the park as both an outdoor classroom and living laboratory where park visitors could learn ways that polluted river water was cleansed through biological processes. They also wanted to create a park experience where visitors could safely play in cleansed water. The design team saw the project as a means of creating an uplifting outdoor experience that celebrated the city's massive modernization project while simultaneously raising environmental awareness (B. Damon 2006, pers. comm., 12 September).

The initial design concept for LWP was reviewed by city officials at the end of four weeks. The design review committee, comprised of Zhang and local officials, was particularly pleased by the park design's resemblance to a fish in the plan drawing view. Fish symbolism was important given the river was once the city's lifeblood; and it also symbolized good health and prosperity. The committee members were also somewhat disappointed by the simplicity of the park's design. It didn't appear to meet their monumental vision of the park as a grand gesture to the city's larger modernization project. Eventually, the city of Chengdu's design team incorporated a larger cultural building that housed exhibitions for environmental education, an outdoor amphitheater and stage, designated children's play area, and an artificially constructed vegetated mountain symbolic of nearby Mt. Emei, *Emeishan*, one of China's four Sacred Buddhist mountains. As noted in an earlier chapter, artificially made mountains and rock piling are part of the Chinese Picturesque design grammar.

Damon (2 May 2019) recently reflected that the collaboration process with Ruddick and the city was authentic where everyone's expertise was considered equitably. For example, Damon acknowledged Ruddick's expertise on natural systems and relied on her to determine the requisite land masses, topography, as well as spatial organization for the various water cleansing stages. The design team agreed on the importance of the impact of the water in LWP: visually, experientially and the river as the source of the park design's water cleansing theme. Damon envisioned the physical link as celebratory and the first appearance of water in the park as a dramatic experience for visitors. Hence, the design incorporated the water drop fountain within a circular settling pond (Figure 5.2) in an arrival plaza with a teahouse. The teahouse building housed mechanical equipment to pump river water up to supply the fountain (Padua 2004b).

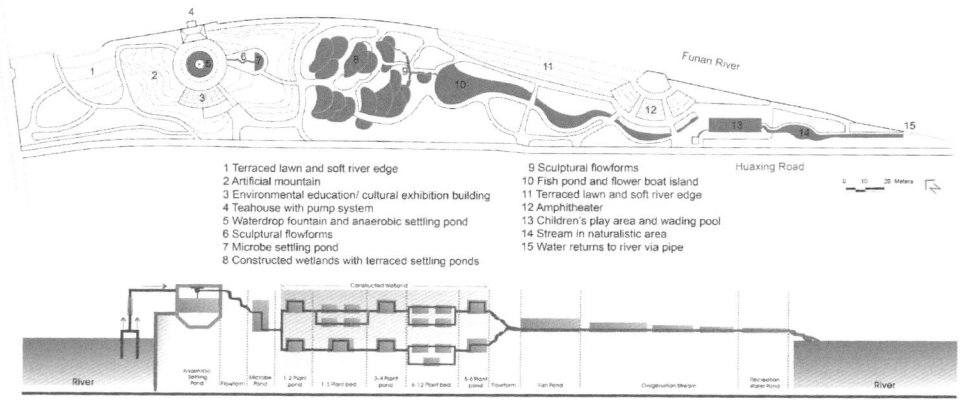

Figure 5.2 LWP Plan and Section based on original, published with permission, graphic by X. Liu

Another important design factor was the relationship of the park's edge to the river. Chengdu's modernization project involved new public infrastructure and the consolidation of new water and sewer lines along the river corridor with its edges defined by concrete walls. Damon discovered these concrete riverfront edges at the LWP site when she returned in 1996 and proposed replacing them with softer edges. After several discussions with local officials, the design team's proposal for vegetated terraces along segments of the park's riverfront edge was accepted. The design team noted the important references to China's rice terraces and agrarian heritage to their client. They also shared historic Chengdu photographs that depicted existing stone steps along the riverfront edge for access to this water source. Furthermore, the proposed stepped riverfront edges were designed to deal with the river's tidal fluctuations and flooding during storm surges. As a design feature, the river also functioned as a backdrop, or "borrowed scenery".

As noted earlier, Ruddick designed the water cleansing process and estimated the various areas needed for each stage. Damon introduced flowforms, a series of vortex basins inspired by John Wilkes' research on water rhythms and oxygenation in natural water purification processes. Damon's design of the flowforms' basins was referential to local ecology and Chinese culture: the phoenix, an imperial symbol and a Chinese mythological figure; a leaf from a Gingko tree, Chengdu's city tree; and a lotus plant, one of the eight sacred Buddhist symbols.

The park's design of natural water cleansing processes set the basis for environmental education and the creation of a didactic experience for visitors. In this light, the park's design of the primary pedestrian path system was integrated with the various water treatment stages. Observation areas, places for sitting and interpretative signage were also located at each stage of the water purification process. At the flowform locations, Damon and Ruddick created tactile and interactive experiences that encouraged visitors to touch the water as it moved through the flowforms. Touching the water (Figure 5.3) represented play and enjoyment, as well as restorative, therapeutic and sensory experiences.

Figure 5.3 Spontaneous interactive experience with living water at flowform, photograph by author

The environmental education/cultural exhibition and teahouse buildings were designed by Chengdu's Architecture and Landscape Architecture Bureaus. Acknowledging the water cleansing process as the park's dominant design feature, the city's designs of architectural components intended to blend into the park's landscape. For example, perception of the building mass of the curvilinear environmental education building was decreased with half of the building facade integrated into the side of an artificially created mountain. The building's primary façade was oriented to Huaxing Road at its lower end with the building's back and side facades set into the slope of the park's artificial mountain. The concept for the artificial mountain was functional, symbolic and aesthetic. In addition to its function as a building edge for the environmental education center and strategy to mask the building's volume, the city's design team understood the mountain's height provided another important function. It created basic gravity for water mobility through the park's cleansing system. The artificial mountain also functioned as both a noise and visual buffer for traffic along Huaxing Road and symbolized nearby Mt. Emei, one of China's four sacred Buddhist mountains.

Chengdu's landscape architects made critical and significant design contributions to LWP's final design and acknowledged the park's biomimetic water cleansing process as the dominant visitors' experience. As noted in an earlier chapter, scene-making and scenery manipulation are foundational to the Chinese Picturesque garden design genre. In this light, Chengdu's landscape architects designated each stage of the water treatment process and various park locations as garden scenes. For example, the locations of flowforms and water bodies were considered scenic waterscapes around which viewpoints and scenes were formulated. They also added features, like the flower island (Figure 5.4), to one of the ponds. Chengdu's designers manipulated park scenes

Figure 5.4 Local design bureau adds flower island to pond, photograph by author

to create the perception of a larger spatial experience. "Twists and turns" and selective plantings were incorporated into the design of the pedestrian paths in the passive and naturalistic areas in the park. Boulders defined informal waterfront sitting areas and large areas near larger water ponds were designed for groups of visitors to enjoy the waterscapes. Riverfront plantings in the park accentuated borrowing the scenery of the river. Additionally, plants indigenous to Mt. Emei were incorporated in the artificial mountain's planting design and the author observed local residents harvesting edible and medicinal plants during field research.

After park visitors walk by or through the environmental education building, they are introduced to LWP's first stage of the water cleansing process: a fountain at the center of a circular plaza. The raised circular fountain is sculpted in the form of a water drop with a single geyser at its center and sits in LWP's circular anaerobic settling pond (Figure 5.5). Damon's fountain design embodies both Chinese tradition and Western European symbolism. In China, the circle symbolizes heaven, and as "harmony" in the *yin-yang* dualism philosophy. The fountain's form is referential to western symbolism of the circle of life and the *"jeux d'eau"*, or visual "water-play" found in fountains of western Renaissance gardens. The water in the fountain and settling pond is supplied from the adjacent river through a mechanical system housed in a teahouse that fronts both the river and plaza. The design of the three-level teahouse building is functional and non-descript. It houses the mechanical equipment that pumps polluted water from the river and it contains indoor facilities for a teahouse and small eatery. Given safety concerns (the anaerobic pond contained polluted river water), Chengdu's design team built a low screen wall as a security measure for the visitors. The wall's height was allowed park visitors to enjoy views of the water fountain sculpted in local granite.

Figure 5.5 Waterdrop fountain and first anaerobic settling pond, photograph by author

Acknowledging Chengdu's teahouse culture, the circular plaza around the fountain was designed as a place for socializing and accommodated informal areas for sitting. This social area has a casual, almost festive atmosphere on a typical weekend. Daily, the sound of the fountain geyser splashing into the settling pond combined with the social activities of tea drinkers creates a vibrant milieu. This plaza area is the only location in the park where tea and food are served.

The visitors' sensory experience from the socially active teahouse plaza area to the second water treatment stage is dominated by the sights and sounds of water moving through the sinuous alignment of custom-made sculptural flowforms. Water swirls within each basin and flows from basin to basin as it oxygenates and descends into a settling pond. The stone path and lush vegetation surrounding this curvilinear alignment of organic-looking elements contain strategically placed boulders for sitting. This novel immersive experience and unrestricted access to the flowforms encourages visitors to touch the water, adding to the intense sensory and restorative experience.

The journey to the third stage, LWP's largest water treatment area and constructed wetlands, consists of a path under and through a vine-covered trellis that descends a steep and wooded slope (Figure 5.6). The combination of the narrow enclosure, deep shade, and steep path dropping nine meters in a short distance creates a sense of mystery and anticipation. For Ruddick, the feeling of this particular zone depicts the magical qualities of "falling down the rabbit hole" in *Alice in Wonderland* (Padua 2004b). At the bottom of the hill, the steep stepping stone path

Figure 5.6 Covered trellis and path descends to the next cleansing stage, photograph by author

opens onto a woodland area. The path then "twists and turns" through a Chinese Picturesque landscape and opens onto the constructed wetlands area via a wooden boardwalk path system.

The path's material change from stone to a wooden boardwalk marks a change of scenery. The width of the wooden boardwalk surrounded by low plantings is three times the width of the stone path. This creates a change from the dappled light cast in the shaded woodland to a sense of openness created by the light from the open sky. The combination of the path change of width and material, and the forest clearing announce the next treatment stage – the constructed wetlands, the largest water cleansing area with various terraced ponds (Figure 5.7).

In design terms, the material changes and the clearing or openness to the sky announce a change in the visitors' pedestrian experience. Wooden benches matching the simplistic design of the boardwalk path are located at the center of the boardwalk path and allow visitors a place to rest before entering the constructed wetlands or main "lungs" of the park. The wooden boardwalk path system leading into and within the constructed wetlands has a random-like quality. No clear directional signage is provided and the visitor is left to wander up, down, and around the terraced curvilinear settling ponds. The boardwalk width varies from two to six meters. The widths appear to be coordinated with the size of the settling pond. For example, five-meter wide paths occur at the perimeter of the larger ponds, with narrow paths located along the small ponds. The varied widths of the wooden boardwalk path system along the

Figure 5.7 Partial view of constructed wetlands and terraced ponds, photograph by author

curves of the ponds create a comfortable scale for observing the biomimetic natural processes.

The system of terraced ponds contains organic matter, water in the process of being cleansed and indigenous wetland plants, and make up the design of the constructed wetlands. Park visitors in this area are surrounded by a mélange of biological water purification processes occurring in a system of over sixteen terraced ponds enclosed by vegetation. These terraced ponds were designed at different elevations or heights, allowing water to filter and flow. The spatial forms of the terraced ponds have an organic quality, yet appear highly constructed. Damon noted that the spatial forms of the terraced wetland ponds were inspired by the travertine ponds at Huanglong National Scenic Area, located in a mountainous region northwest of Chengdu.

The treated water from the constructed wetlands flow into another series of flowforms and traverse into another settling pond enclosed by dense vegetation. From this pond, the water is then treated through another system of flowforms before traversing into LWP's largest water body (Figure 5.8) containing fish. These live fish were indicators that this water body met China's category III environmental quality standard and was considered safe for swimming and fishing. Purified water from this large fish pond flowed downstream through an articulated stream corridor and into a wading pool at

Figure 5.8 Largest water body containing fish, photograph by author

the designated children's play area. The wading pool marks the children's play area and sits within a field of stone paving.

The bottom end or southeastern end of the park suffers by comparison to other locations in the park. The amphitheater and wading pool in the children's play area fight against the integrity of the design rather than complement it. These hard-surfaced areas virtually span the distance from river to city edge, cutting off the bottom of the park. Beyond the children's play area, no changes were made to the original design, and cleansed water channels into a narrow stream within a lushly vegetated naturalistic area before disappearing into an inlet. The cleansed water discharges from the pipe inlet into the river. At this end of the park, the path is informal and unimproved. There is no signage or place marker to signal the end of the path or water cleansing process. The only way to exit the park is to retrace the path back to the body of the park. The construction of the park took less than a year. Its completion and formal opening coincided with a major national holiday on the first of October in 1998.

At a later field trip to LWP, the author observed the amphitheater's hardscape was "softened" by changes to the original plan by the city's designers (Figures 5.9 and 5.10). Terraced lawn replaced some of the benches and steps in paved areas. These changes suggest the city took seriously the author's written critique of the amphitheater's original design in a published essay (Padua 2004b).

Figure 5.9 LWP Amphitheater's dominant field of paving in 2003, photograph by author

Figure 5.10 LWP Amphitheater's paving is softened with vegetation a few years later, photograph by author

Hybrid modern synthesis

Chengdu's Living Water Park exhibits strong influences from international trends in landscape architecture; but, it is also distinctively Chinese. It is the first ecologically oriented park of its type in China, and it was unusual in its didactic character at the time it was built. Neither of these concepts would have been alien to public parks in the west. However, it represented major departures from Chinese park design traditions in the late 1990s. Simultaneously, LWP's final design execution incorporated many elements of the Chinese Picturesque tradition, e.g. artificial mountain and rock-piling, scholar rocks (Figure 5.11), paths that "twist and turn", as well as scenery manipulation. More importantly, LWP renewed the importance of the river to the city's identity, as well as local cultural heritage, and commemorated the city's larger modernization project. Simultaneously, Chengdu demonstrated leadership in terms of the city's sustainable nation-building efforts to restore the river corridor and arrest environmental degradation.

As noted earlier, the closest analogue to LWP is Lorna Jordan's Waterworks Garden in Reston, Washington, south of Seattle. Jordan's project was completed in 1996, two years before the opening of Living Water Park. Both parks were conceived by environmental artists and reflect common concerns about the sacredness of water. Visitors at both sites experience a progression of the water cleansing process. Jordan's project treats water from a waste water reclamation plant in a suburban area and is only accessible via private car. In contrast, LWP is located in a densely populated urban area and is accessible by foot, as well as public transit or private car. It demonstrates cleaning the adjacent polluted Funan river water through a series of natural processes

Figure 5.11 Scholar rock with park name engraved at LWP entry, photograph by author

in different areas within the park before returning it to the river. The designers also incorporated interpretative signage with text that explained stages of the cleansing process (Figure 5.12). Jordan purposefully did not utilize interpretative signage in her Washington park (Gonzalez 1998).

Much of LWP's design is international in concept and style. Elements such as the waterdrop fountain or sculpted flowforms may contain references to Chinese culture and history. However, these would not be recognized as representative of the Chinese Picturesque genre (Figure 5.13). In many ways, the Chinese references that inspired designs of the flowform basins can be interpreted as a romantic and foreign view of local design. The water cleansing process dominates LWP's design and creates a highly sensory and immersive visitor experience. Chengdu's landscape architects added elements of the Chinese Picturesque to the original LWP design. In fact, the park's main entry sign defines the main features of the water cleansing process as a scene, for example, "forest walk", "mountain forest walk" and "flower walk" (Figure 5.14).

LWP is the first award-winning park in post-Mao China to have exhibited this combination of international design influences, the Chinese Picturesque genre, local identity and nation-building. It was the first time Chengdu's various stakeholders rallied around environmental issues as the key to renewing the city's local identity. At the time, Chengdu was a national leader for environmental stewardship with their heroic

Figure 5.12 Interpretative signage at first set of flowforms added years later, photograph by author

Figure 5.13 Sculptural flowforms atypical of Chinese Picturesque, photograph by author

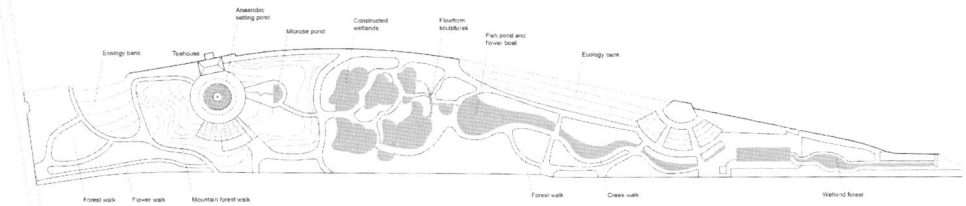

Figure 5.14 Scenes called out in LWP signage, based on original, graphic by X. Liu

river clean-up and modernization efforts. LWP also represents ways that designers of purpose-built parks were allowed to innovate in secondary cities during China's late 20th century. International design influences, particularly ecological design and biomimetic water cleansing processes, as well as environmental art, were imported via Damon and Ruddick, both based in the USA. The Chinese Picturesque tradition was carefully incorporated by the city's designers. These various dimensions, local identity, nation-building and the fusion of international design influences with the Chinese Picturesque represented in LWP's spatial forms contribute to the synthesis of hybrid modernity. A new design language with accompanying grammar and vocabulary is illustrated in Table 5.1 and is somewhat self-explanatory, with design approaches called out with related design grammar, vocabulary, and material expression.

The duality of the project's design went deeper than just the combination of different elements drawn from different traditions. In many cases, specific design elements meant different things to local designers, local residents and foreigners. Damon and Ruddick saw many elements of the design as having a clear Chinese quality: ginkgo leaf, lotus plant and the phoenix, the circular water drop fountain, the references to Mt. Emei or the Huanglong limestone terraced ponds, and the plan view of a fish form. Local designers and officials saw these as innovative foreign ideas or else invested them with different local meanings – such as conformity to the traditions of rock piling or borrowing scenery. The local community visited the park for leisure purposes (Figure 5.15) and its novelty. LWP at the time was considered China's first ecological park and was on the official list of special projects that government and party officials were mandated to visit and learn from.

The outcome of this process was the first purpose-built park in China that represents hybrid modernity. The city was focused on reconnecting with the heritage of its "local" river identity through environmental remediation and LWP park commemorated this effort. It represents a creative collision of Chinese Picturesque language with international design influences and remakes the city's identity. The timing of LWP's completion was critical. It opened at a time of escalating environmental concern in China and rapidly became a flagship project, both for its environmental focus and didactic character. It was a standard destination for mayors from all parts of China to visit, and hundreds of municipal officials consider it an exemplar and aspirational project to have in their cities. In this light, the project contributes to nation-building. LWP has won major awards inside and outside China, and virtually every young Chinese landscape architect, as well as international students, studied the park at the time. All of this has combined to give Living Water Park a seminal role in the contemporary

Table 5.1 Summary of case study one: LWP's Hybrid Modern design language, graphic by author

Design Approach	Design Grammar	Design Vocabulary	Materiality
Environmental Activism + Ecological Design	Commemorate Chengdu's modernization project Celebrate living water, natural systems and local ecology	Water cleansing via biological processes Biomimicry and constructed nature Organic forms	Aerobic, anaerobic matter Indigenous plants Settling ponds, constructed wetlands, new habitats Local stone In situ concrete
Didactic	Water treatment and riparian ecology Commemorate river clean-up Environmental education	Pedestrian path system integrated with water treatment stages Flowforms enable sensory experiences: visual, tactile and auditory Interpretative signage	Stone Aggregate and cobblestone paving Boulders for sitting Stone for custom-made flowforms Flowing water Live fish –water quality indicator Stone & wood for signage
Cultural Local identity	Urban regeneration Local history Symbolism	Everyday riverfront use Huanglong Ponds Mt. Emei River as city icon River, aesthetic topic for poets Agrarian heritage: rice terraces Fish Circle (Chinese + European)	Stone steps Riverfront promenade Spatial forms for constructed wetlands Engineered fill for artificial mountain Native plants and unrestricted access to lawn In-situ concrete/stone Terraced riverbanks Plan interpretation Water drop, heaven
Nation-building	Leaders of environmental movement		First urban ecological park in China and the world
Chinese Picturesque	Scene-making Scene-manipulation	Waterscape, woodland Borrowing scenes Framing scenes Twists and turns	Stages of treating water Circulation and path design, viewing while moving, viewing while stationary Expands spatial depth of field Spatial enclosure created with natural (mineral, vegetal and water) materials

history of landscape architecture in China. It also contributes to landscape urbanism and the later ecological urbanism, as well as landscape systems and blue-green infrastructure design. It was designed as a landmark and commemorative to the city's modernization project, and it is the first exemplar of China's late 20th century hybrid modernity.

Figure 5.15 Informal gathering area along LWP's fish pond edge, photograph by author

Case study two: Zhongshan Shipyard Park, *Zhongshan Shiqi Jiang Gongyuan*, Zhongshan, Guangdong Province

"Landscape design is a process that visualizes the meaning of the site through preserving and modifying the existing forms, and if necessary, creating new forms".

Yu, K. p. 36 (2003)

Zhongshan Shipyard Park, known locally as *Qijiang Gongyuan*, is located in the city of Zhongshan, Guangdong Province in southern China and was completed in 2001. The park covers eleven hectares (twenty-seven acres) and sits on a former industrial site along Shiqi River, a tributary of the Pearl River, *Zhu Jiang* (Figure 5.16). It was designed by Turenscape, China's first private firm (not a state-owned enterprise) offering professional services in landscape architecture. Yu Kongjian, Harvard-trained and a native of China, established the Beijing-based firm when he was then Director of the Center for Landscape Planning and Design at Peking University. He currently is Dean at Peking University's College of Architecture and Landscape Architecture and Turenscape President. This purpose-built park was designed to celebrate China's post-revolutionary industrial heritage – the period when Mao ruled a country largely organized into communal factories or "work units", known as *danwei*. This is an era that the Chinese population in the 1990s was eager to forget; although it was a watershed period in the creation of the People's Republic of China.

Pearl River Delta

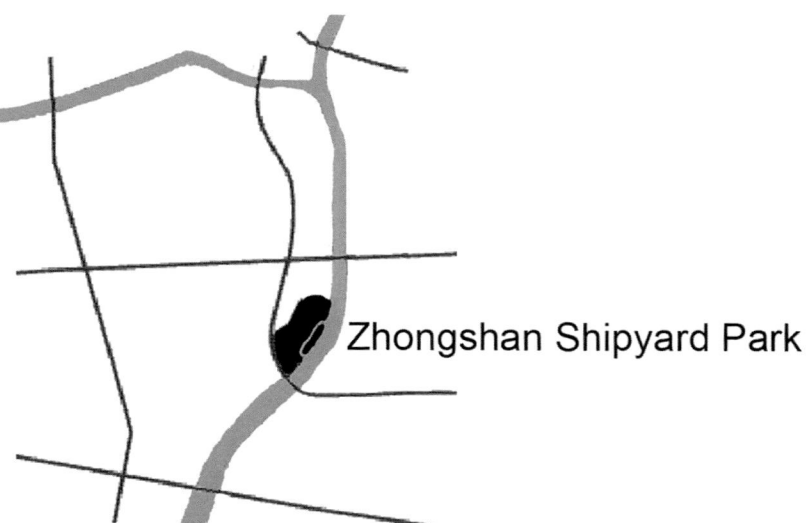

Zhongshan Shipyard Park

Figure 5.16 Context and location map for Zhongshan Shipyard Park, graphic by X. Liu

As principal designer for Zhongshan Shipyard Park (ZSP), Yu played a key role. ZSP represents a breakthrough design for landscape architecture in China. It was the first public park in China to commemorate Mao's industrial legacy and recycle the shipyard's artifacts. It received numerous international design awards and launched Yu's professional career. Aspects of the park's design reflect Yu's vision of landscape architecture in China, and he is now a major voice in China and internationally. ZSP's design is simultaneously very strongly influenced by contemporary western design and deeply Chinese in its homage to the revolutionary history of the nation. Furthermore, ZSP's design also deploys Chinese Picturesque design grammar.

Zhongshan context

ZSP is located on the southern edge of this secondary city's urban core. Zhongshan is situated along the western edge of the Pearl River Delta (PRD) in Guangdong Province (Figure 5.16). The city is eighty-six kilometers (54 miles) south of Guangzhou, the provincial capital, and northwest of the Hong Kong (HK) Special Administrative Region (40-minute ferry ride from HK). Zhongshan is one of four cities and twelve counties in the 1985 PRD Economic Open Zone designated by the central government. These PRD locations were designated for so-called Foreign Direct Investment (FDI) and large-scale knowledge transfer (Bie et al. 2015). ZSP's location on Shiqi River, a tributary of the Pearl River, China's third longest river, made it a natural site for the Yuezhong Shipyard.

Zhongshan's urban history, like that of many established cities in China, is very long. Its development is informed by its delta location and reflects the evolution of southern China and the nation. A brief summary informs its significance as a city where a breakthrough park design came to light. It was first known as Xiangshan and emerged in the 11th century as a fortified market town. Its prosperity was based on major salt production. Many imperial court officials and family members were known to have fled here after the downfall of the Southern Song dynasty circa late 1270s (Ye et al. 1990). Later, its development grew as a result of its proximity to Guangzhou, formerly Canton, and China's establishment of the Canton System circa 1757 with the subsequent long period of overseas trade and western influence. The First Opium War circa 1839 took place in the area and as the city and region grew open to western influence, its youth studied abroad, including Dr. Sun Yat-sen (in *Putonghua Zhongshan*, Yat-sen is a Cantonese dialect). It was renamed Zhongshan in the 1980s for its most prominent native son. As noted in an earlier chapter, Sun is deeply revered throughout China for his role in the 1911 Republican revolution that caused the downfall of hundreds of years of imperial rule. The father of modern China and Zhongshan's native son, Sun is cherished in southern China, and is an important element of the city's identity. This association also reinforces the city's self-image as a progressive place for innovation.

Project inception

When Yuezhong Shipyard was operational, it was a state-owned-enterprise or communal factory that manufactured and repaired ships, and housed the workers (Figure 5.17). These collective work units were one of the foundations of Mao's communist society and were more than just employment centers. These were monolithic social institutions that took in all aspects of the workers' lives. Shelter was provided through onsite dormitories for workers along with canteens where they ate. Children were educated onsite, and families relied on the rudimentary recreational areas within the factory compound for leisure activities. All aspects of social welfare, including health care and retirement, were provided by the collective. Employment involved a type of cradle-to-grave security that was known as the "iron rice bowl" (Padua 2003).

Yuezhong Shipyard was the city's industrial heart for nearly half a century and Zhongshan's largest employer from its founding in 1953 to the early 1980s when employment peaked at 2000 workers. It finally closed in the late 1990s after declaring

Figure 5.17 Yuezhong Shipyard factory circa early 1990s and park site, photograph provided
        by Turenscape, published with permission

bankruptcy. Prior to that time, the shipyard had an unbroken history that spanned
the 1950s Great Leap Forward, the upheavals of the 1966–1976 Cultural Revolution,
Deng's reforms and related post-Mao rapid growth of export manufacturing (Padua
2003).

In addition to a review of available literature and archival records, the case study
analysis of ZSP involved interviews with Deputy Mayor Peng Jiangweng (client), Yu
Kongjian and Pang Wei from Turenscape, and interviews with other design educators.
Like work on LWP and the other case studies, field research was critical to understanding design intentions and the synthesis of hybrid modernity.

In the mid-1990s, Zhongshan city officials observed the shipyard's ultimate
decline. This area of southern China experienced extremely rapid growth in less than
three decades. Zhonghsan's PRD location was due west of Shenzhen, China's first
Special Economic Zone, a former fishing village that rapidly urbanized as a result of
its physical and economic relationship to Hong Kong. The PRD region was home to
the greatest concentration of the manufacturing industry and gained the late 20th century reputation as "workshop of the world". If it were a separate nation in the 1990s,
its economy would have ranked among the twenty largest in the world. Guangdong
Province's manufacturing industry at the time was largely devoted to export products,
such as electronics and consumer goods, and Zhongshan was one of these manufacturing centers. By the late 1990s, the shipyard was deemed economically unsustain-
able; and it was kept alive with infusions of money from the state banks – loans that
would never be repaid. Zhongshan's Deputy Mayor Peng was tasked to find new

uses for the riverfront industrial site, especially given it occupied a strategic location between Zhongshan's commercial core and residential areas. It could not be allowed to become derelict without serious implications for urban well-being, especially the city's image.

Zhongshan officials believed the establishment of a new public park would stimulate interests in property development in the central core and promote urban regeneration. They also recognized that serious environmental degradation had taken place in the region during the preceding twenty years, and one of their goals was to strike a balance between promoting industry and creating a habitable city. They were encouraged by past efforts when Zhongshan received designation in 1996 as a Chinese National Garden City, a biennial award given by the Ministry of Construction (Padua 2003). Furthermore, Zhongshan's Mayor Huang Ziqiang was awarded the 1997 United Nations Habitat Scroll of Honor, a prestigious international award for outstanding contributions to improve the urban environment.

Peng met Yu at a Mayors' Forum in Beijing in 1998. At that point, the Yuezhong Shipyard was already slated for closing. They discussed a variety of issues related to urban renewal at the conference, but not the design for the shipyard's replacement. Zhongshan officials had already gone through the typical design competition process and were prepared to hire a French designer. After returning from the Mayors' Forum, Peng pressed other city officials to consider giving the design commission to Yu and Turenscape, a China-based private company. Eventually, a meeting at the shipyard site was arranged for Yu and Zhongshan officials. Subsequently, the city commissioned Turenscape on the condition that design approval could be achieved through their review process. The city had tasked a Design Review Commission comprised of senior Chinese landscape educators and local officials. Turenscape's design was approved after an unusually long (nearly one year) review period in China and construction commenced expeditiously. The park was completed in May 2001(Padua 2003).

The city's vision was focused on the shipyard's strategic location as a significant opportunity for urban regeneration and renewed identity. A successful, high-profile purpose-built new public park would enhance the city's overall image as livable. Hence, remaking the city would help efforts to attract FDI and spur the growth of high-technology and related manufacturing industries. The science and technology sector was considered key to China's economic reforms and FDI allowed for joint development, advanced knowledge and technology transfer. Locations within the PRD Economic Open Zone status received higher priority in this period. Within this light, the city's urban regeneration goals for the new public park on the former industrial site focused on the city's downtown identity, real estate investment opportunities, improving recreational opportunities and community well-being, and creating a regional tourist destination. City officials also sought to create visible signals that addressed the community's concerns for the environment and sustainable future.

The Pearl River carries the second largest water volume after the Yangtze River. Various cities and counties in the PRD were still in the process of modernizing sewage and water treatment infrastructure in the 1990s and simultaneously dealing with pollutants in industrial wastewater discharging into the delta. Remediating environmental degradation contributed to the city's leadership concerns for the city's well-being. Poor water and air quality, and shortages of electricity in the region, were critical issues that received a great deal of publicity at that time and were a source of considerable discontent. Like the citizens of Chengdu, Zhongshan's residents at that time considered environmental issues a high priority.

Turenscape developed a variety of aesthetic and social objectives that helped shape their approach for the design of Zhongshan's new purpose-built park. Their intention was to serve the client's interests and create an innovative design that was sensitive to the environment, site's history, local culture and surrounding physical context. Yu, Turenscape's principal designer, also had the unspoken goal that exists in their many projects – to create an award-winning design and break new ground in landscape architecture (Padua 2003). Their strategy went against the grain of the modern notion of "*tabula rasa*" where designers would clear the site, erase its memory and mimic the western neo-classical or Chinese Picturesque design genres.

Turenscape followed the *genius loci* design approach and developed a design that celebrated the spirit of place – the site's industrial heritage and natural resources. Yu and his team reimagined the site's natural ecology, as well as its cultural history, especially Mao's industrial era. Their approach honored the factory workers who had toiled and lived there by creating a design that memorialized their work and everyday life. Yu, in particular, saw the need to create a design that could engage the public not only in the history of the site but also the recent history of the Chinese nation. At the time, it was a challenge given the majority of the Chinese public had developed a prepossession with change and modern urban life and turned its back on the recent history of the Mao era.

Simultaneously, for some of China's educated population, nostalgia for traditional China and a romanticized Confucian society had surfaced (W. Wang 2009, pers. comm., 12 June). Yu, a staunch CPC member, believes the imperial era was a backward time – a major disruption for China's road to modernity. For Yu, progress in design thinking among educators and professionals in China's discipline of landscape architecture had been oppressed by the Chinese Picturesque genre, a symbol of imperial China.

Turenscape interviewed former shipyard workers to understand everyday life at the shipyard. These oral histories were an important part of Turenscape's design inspiration. Turenscape also met with the city's flood control engineers to assess issues posed by the site's riverfront location. Their site research informed Turenscape's comprehensive understanding of the shipyard's built form and materials, its industrial use and existing natural setting and vegetation, dynamics of the river and tidal fluctuations, visual quality, the site's relationship to its urban context, as well as the everyday uses by the workers. In Yu's words, the main objective was "to artistically and ecologically dramatize the site's spirit" (Padua 2003). In effect, Yu combined the *genius loci* or "spirit of place" with ecological design. Yu and the Turenscape team deployed three basic design principles, "preserve, re-use, and recycle", with goals to retain existing vegetation, natural habitats, water features, and historical and industrial elements. It also involved the adaptive re-use of existing structures, materials and forms for new functions, as well as the creation of new forms for new uses. Additionally, Turenscape's research involved research on two international design precedents for parks located at former industrial sites or so-called brownfields: 1) Gas Works Park in Seattle, Washington, designed by American landscape architect Richard Haag (1923–2018) and 2) Landschaftspark Duisburg Nord, or Duisburg Park in Germany, designed by German landscape architect Peter Latz.

Gas Works Park covers over 20 acres (8 hectares) on a post-industrial waterfront site and was completed in 1975. It occupies the site of the former Seattle Gas Light Company gasification plant built circa 1906, closed in 1956 and later purchased by the city of Seattle with the goal to convert it into a public park. Haag's 1971 design of Gas Works Park was considered a ground-breaking and innovative park design that transcended design approaches within the discipline of landscape architecture at that time (Way 2015). It is known as America's first public park that accomplished two major design ideas: 1) preserved the massive building complex of a disused gas plant and the site's industrial heritage; and 2) reclaimed polluted soils in the toxic park site through biological processes. Haag's design also called for the conversion of some of the existing industrial structures for recreational uses. Haag's design was an homage to America's industrial energy-producing heritage, given it was the sole remaining coal gasification plant in the United States. His park design combined the significance of both cultural heritage and ecological systems. An award-winning project, Gas Works Park has also been construed as the post-industrial archetype for ecological urbanism in North America (Way 2015).

Duisburg Nord Landscape Park was designed in 1990 by Peter Latz, a German landscape architect. It covers 230 (570 acres) on a former industrial site that contained the Thyssen Steelworks manufacturing plant. Located in Germany's Ruhr area, formerly known as Europe's coal-mining and steel manufacturing center and Germany's industrial heartland in the 19th and 20th centuries, Latz's park was part of the larger 80-kilometer (50-mile) long Emscher Park. This larger park was designated in the 1989–1999 International Building Exposition (IBE), *Internationale Bauausstellung*, a ten-year program of exemplary and innovative integrated projects spearheaded by Land Northrhine-Westfalia, one of Germany's western states (Stilgenbauer 2005). This program was responding to the area's economic decline and environmental degradation due to its industrial heritage. It focused on environmental protection, nature conservation, preservation of the area's industrial heritage, creation of tourist attractions through the adaptive re-use of industrial buildings and cooperation in the region's planning and administration (Schepelmann et al. 2016).

Latz's 1990 park design was realized in phases between 1990 and 2002. Like Haag's project, it is considered an exemplar park design on a post-industrial site. Larger than Haag's project, and more complex in terms of the fragmented and sprawling aspects of the industrial plant site, Latz's design combined social, cultural and ecological concerns. After careful site analysis, Latz's design approach involved utilizing the inherent spatial forms created by the sprawling industrial building complex, existing patches of vegetation and natural resources – the site's "spirit of place". In terms of realization, Latz deployed a park systems strategy that integrated the functions of various independent design components. These were intended to give visitors a sense of location and orientation in the park, as well as create a human-scale experience (Stilgenbauer 2005). To meet the goals of creating recreational opportunities and attract park visitors, Latz incorporated various design innovations including the conversion of an existing large diameter gas tank into a scuba diving center; and the re-use of the exterior walls of concrete bunkers for rock climbing. Latz's park design is a series of smaller parks nested within the former industrial site and connected by pedestrian paths, elevated promenades and bicycling paths. It also includes a large public space for assembly and cultural events. Latz also considered ecological succession as a landscape systems strategy where vegetation and natural systems would eventually take over and visually dominate the park's industrial heritage. With attention to natural systems, ecology and the environment, the park creates a didactic experience for environmental education geared to all age groups, and visitors are introduced to ways the site's industrial heritage can be preserved.

Yu and Turenscape worked with the city officials' regeneration objectives and laid the groundwork for their design approach in initial client meetings. They presented their findings from their analysis and research on Haag's Gas Works Park and Latz's Duisburg Park. Gas Works Park was opened to the public in the 1970s and early phases of Duisburg Nord Park opened to the public in 1994. Turenscape's studies of these reclaimed brownfields and post-industrial park designs reinforced their design approach to preserve the spirit of place – the site's industrial and cultural heritage as a shipyard and Mao era collective work unit. Turenscape also presented the opportunities afforded by the site's natural resources: existing water features and mature vegetation, as well as the major challenges of flood control and the city's intention to widen the river along the park site's boundary.

For Yu, the Yuezhong shipyard was emblematic of Mao's industrial period and the post-Mao New Era created by Deng's reforms for both Zhongshan and the nation. Over half a century, the Yuezhong shipyard workers had lived through the Great Leap Forward, Mao's Cultural Revolution, Deng's reforms and related export boom, along with China's entry into the world economy. Yu noted "the project is a small site with a big story" (Padua 2003). The combination of the design review and community feedback process was unusually long for a park design in China. Typically, a public park is designed by the city's local design bureau and built by their local construction bureau without any oversight. Turenscape's general design concept to celebrate the shipyard's industrial heritage and recycle existing industrial machinery was considered controversial and prolonged the government review process. Live design presentations by Turenscape were televised locally and physical models were displayed in city government buildings (Padua 2003). Feedback from the public was garnered through the use of written comments left at "suggestion boxes" located in buildings where physical models of the design proposal were displayed. Most of the feedback was positive, given the local community had family members who worked in the shipyard and were pleased by the idea that the park would serve as a memorial to their lives.

Design realization

As noted earlier, Turenscape's design intentions were based on commemorating the site's spirit of place and the nation's industrial heritage. However, their initial concept spatially organized the new park using the Daoist dualist principle and *yin-yang* symbol for harmony (Figure 5.18). The team designated the site's northern area, the "active" (yang) or urban zone. The majority of the shipyard buildings had also occupied this area of the site. The southern part beyond the existing lake where vegetation and mature trees existed was designated as the "passive" (yin) zone. Within this *yin-yang* or passive/active context, the park's program overall was not complex. It incorporated active areas where all age groups could gather, play and socialize and spend leisure time, and a new cultural building containing a teahouse, exhibition spaces,

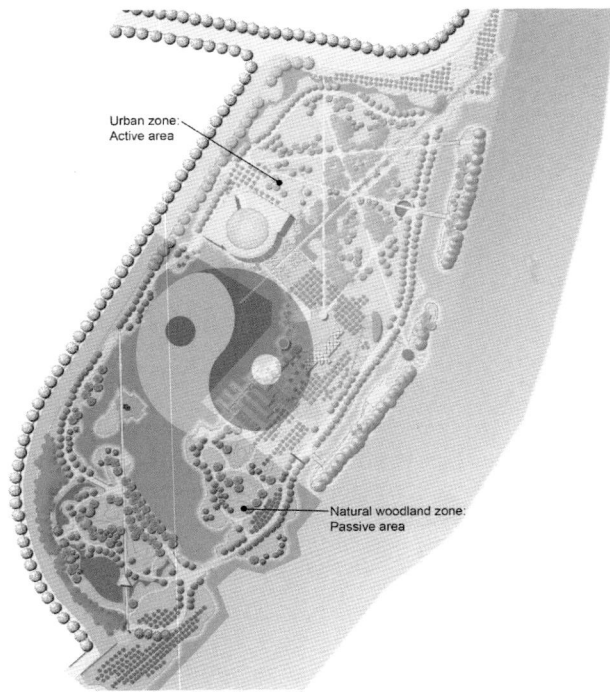

Figure 5.18 Turenscape deploys the Daoist yin-yang dualist principle for harmony when organizing the park's design, graphic by Turenscape, published with permission

park administrative and maintenance offices. Passive areas in the park contained informal places to sit, socialize and have picnics, and enjoy views of the lake.

ZSP's design of active and passive zones was not complex. However, the park's spatial organization did represent a type of design complexity unprecedented for China at the time. Inspired by the site's spatial forms, especially the dominance of the existing rectilinear buildings and linear railway, Turenscape derived a system of axial geometry. Similar to Beaux Arts and renaissance garden traditions, Turenscape's deployment of the design axis was a type of organizing design principle that defined sight lines, visual corridors and focal points, and path movement systems for physical connectivity within ZSP (Figure 5.19).

ZSP's primary design axis, a major pedestrian promenade, was comprised of the re-purposed railway tracks at its original location. As an organizing design device, this railway path creates a visual corridor and pedestrian experience that connects the north entry to the lakefront and introduces visitors to the site's industrial past. The railway path's axis and visual corridor also extends and traverses across the lake to a pavilion on an artificial island. Two other linear axes in the urban zone are defined by two pedestrian paths that depart from the "Red Box", another major design feature and newly constructed design element (Figures 5.20 and 5.21). These two axial paths traverse through the urban zone with industrial artifacts at their visual terminus.

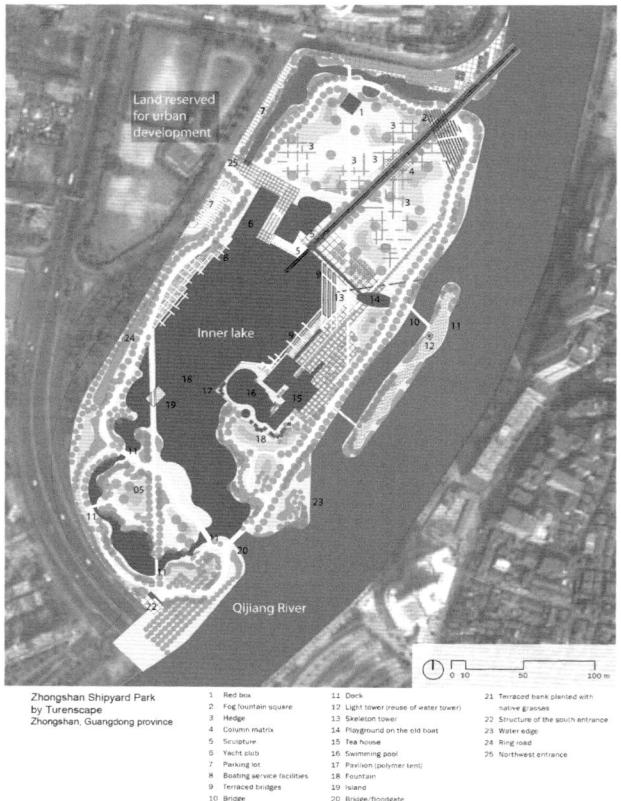

Zhongshan Shipyard Park	1	Red box	11	Dock	21	Terraced bank planted with
by Turenscape	2	Fog fountain square	12	Light tower (reuse of water tower)		native grasses
Zhongshan, Guangdong province	3	Hedge	13	Skeleton tower	22	Structure of the south entrance
	4	Column matrix	14	Playground on the old boat	23	Water edge
	5	Sculpture	15	Tea house	24	Ring road
	6	Yacht club	16	Swimming pool	25	Northwest entrance
	7	Parking lot	17	Pavilion (polymer tent)		
	8	Boating service facilities	18	Fountain		
	9	Terraced bridges	19	Island		
	10	Bridge	20	Bridge/floodgate		

Figure 5.19 ZSP Illustrative Site Plan by Turenscape, published with permission

The Red Box, a red-painted steel enclosure with four walls spanning around forty feet (twelve meters) and open to the sky, commemorates Mao's era and symbolizes the shipyard's contribution to industrialization and self-reliance. Inside the Red Box, the twelve-foot (four-meter) high walls enclose a reflecting pool and create a contemplative experience. Some of the elder citizens interviewed onsite interpret the Red Box as a symbol of Mao's Cultural Revolution, the first time that this dark period was publicly acknowledged. The bifurcated path that leads out of the Red Box traverses a lawn area that contains outdoor garden rooms defined by tall clipped hedges. These garden rooms replicate the size of the dormitory rooms where shipyard workers lived and the designers intended to provide a spatial experience for visitors to learn about life in a collective work unit (Figure 5.22). The garden rooms are often occupied by small groups or individuals enjoying their leisure time. While the park's main path system was linear, lacking in Chinese Picturesque "twists and turns", the paths helped define linkages between the various scenes within the park's active zone.

Turenscape's park design strategically located a series of shipyard scenes throughout the park. Some of the scenes literally captured the site's former industrial heritage with figurative sculpture mimicking daily chores. For example, a bronze sculpture

Figure 5.20 Red Box with bifurcated path and reflecting pools, photograph by author

Figure 5.21 Framed views through Red Box, photograph by author

Figure 5.22 Green hedges define spaces that replicate the lay-out of the shipyard's communal dormitory rooms, photograph by Turenscape, published with permission

depicts two workers with industrial equipment gesturing at a crane attached to one of the large gantries (Figure 5.23). Another bronze figurative sculpture depicts a worker climbing bamboo scaffolding adjacent to a renovated smokestack from the shipyard's foundry. In addition to creating industrial scenes, Turenscape frames lakefront waterscape scenes by stripping building skins from two existing large industrial lakefront structures (Figure 5.24). Both former sheds now function as open-air pavilions with one retaining its former use as a boat dock (Figure 5.25). Both places create stationary points for visitors to enjoy framed views of the lake's waterscape and backdrop of the park's naturalistic vegetation in the passive zone.

ZSP is a breakthrough design in China. It is the first park to reclaim a brownfield site and preserve Mao's industrial heritage. However, it also demonstrates some major ecological design moves never seen before in the Pearl River Delta and China. Turenscape's didactic design reveals the complexity of the lake water's daily tidal fluctuations (over a meter) and its relationship to the adjacent Qijiang River. Turenscape created an immersive environmental educational experience along the park's western lakefront with a path network of stepped platforms in the form of a grid (Figure 5.26). Various native plants were incorporated into the open grid network. With unrestricted physical access to this lakefront and open views, visitors can observe changes in the lake's water levels and learn about the dynamics between the adjacent river and the site's lake environment. It was also the first time native wetland grasses were used in Zhonghsan. Local officials were accustomed to ornamental flower beds, especially given the area's heritage for cultivating flowering Chrysanthemums.

Figure 5.23 Human scale bronze figures of workers capture the shipyard's everyday life, photograph by author

Figure 5.24 Yuezhong Shipyard's lakefront sheds, photograph by Turenscape

Figure 5.25 Former sheds are re-purposed as lakefront pavilions, photograph by author

Figure 5.26 Stepped path system and native plantings create access to the water and an educational experience, photograph by author

Figure 5.27 Ecological Island preserved an existing stand of mature trees and the site's historic riverfront edge, photograph by author

In a more dramatic move for landscape architecture in China, Turenscape preserved an existing stand of over ten mature Banyan (*Ficus microcarpa*) trees through the creation of an artificial island (Figure 5.27). Turenscape convinced the design review committee about the significance of the trees as the site's historic boundary and its natural climate control function for the city's subtropical zone. Instead of massive clearance of the trees and the site's historic riverfront embankment to widen the river, the city's engineers implemented Turenscape's alternative that involved strategic excavation around the trees' roots and the construction of a wall and canal to both preserve the mature trees and manage river flooding. The newly made island, termed "ecological island", incorporated pedestrian paths, shaded sitting areas and re-purposed steel bridges that linked it to the main park (Figure 5.28). Saving the mature trees from river widening by city engineers and creating an artificial island to preserve them *in situ* was considered heroic landscape architecture in China at the time (Z. Bao 2009, pers. comm., 30 April).

Turenscape's philosophy, "preserve, recycle, re-use", was carried throughout the realization of their innovative design. Spatial forms and design ideas inspired by the site's cultural and natural resources depicted in the completed park were considered new in China at the time. These innovations considered design approaches that combined the site's spirit of place, temporality and ecological design. The notion of temporality is expressed by site references to Mao's nation-building efforts and industrial era, the changes represented by the site's conversion from industrial to recreational use and the dynamics of daily tidal fluctuations. This purpose-built park created new experiences for visitors that exceeded typical recreational and leisure activities provided in Chinese Picturesque or western neo-classical parks at the time.

The urban side of the park continues to be heavily used. An artificial creek (Figure 5.29) along the site's northern edge functions as a playground for children

Figure 5.28 Ecological Island contains shaded sitting areas and path, photograph by author

Figure 5.29 Artificial creek functions as a noise buffer and location for water play, photograph by author

Figure 5.30 Allée of Palm trees and stone paving with steel bands that celebrates the park's industrial heritage, photograph by author

and adults to play in the water. It also serves as a noise buffer to the vehicular traffic along the adjacent frontage road. Turenscape incorporated an *allée* of Palm trees set in stone paving near the artificial creek (Figure 5.30). The field stone paving incorporates bands of steel that celebrates the park's industrial past. Parts of the park have a living museum quality with contrived everyday scenes of workers in bronze sculptures and industrial shipyard artifacts located along pedestrian paths. By memorializing the workers with human-scale figurative bronze sculpture, Turenscape depicts them as everyday heroes. Traditionally, figurative sculpture (monumental scale) in parks and public space around China were usually reserved for local heroes, e.g. poet, musician, emperor, or revolutionary leaders. During field research, the author observed visitors photographing themselves with bronze sculptures (Figure 5.31) and in various locations in the park's urban zone. They posed with the bronze figures and took group photos at the railroad path. Wedding photography also takes place in the park, with couples posing in contemporary western fashion (Figure 5.32). In China, the success of a public park is demonstrated when it is utilized as a backdrop for wedding photography.

Hybrid modern synthesis

Zhongshan Shipyard Park represents hybrid modern design. Its spatial forms illustrate the fusion of international design influences and Chinese Picturesque, nation-building and local identity. Yu, Turenscape's President, is a native Chinese, educated in landscape architecture first at Beijing Forestry University then awarded the Doctor of Design after three years of study at Harvard University. Yu learned the Chinese Picturesque tradition in his younger years and has since developed strong feelings

Figure 5.31 Visitors take photographs with bronze sculptures, photograph by author

about its detriment to the progress of landscape architecture in China. Yu's international design influences from his Harvard education were deepened by his two-year experience at the SWA Group in southern California. Zhongshan, a secondary city, with their needs to remake the city's identity in the late 1990s, provided the crucible for Turenscape's foray into innovative park design and the creation of a hybrid modern public park as a city landmark in late 20th century China (Figure 5.33).

The most obvious of these key dimensions in ZSP is the emphasis on local identity. The central and dominant theme of the park's design is the site's history and link to Mao's era. Turenscape went to great lengths to incorporate elements of that past, educate the community about Mao's nation-building efforts, as well as contemplate Mao's dark period of the Cultural Revolution. Retaining and re-purposing the former shipyard's industrial artifacts to create a type of cultural destination were also intended to teach visitors about transformations in Deng's New Era and China as a whole. Introducing immersive experiences to teach visitors about the lake's tidal fluctuations and preserving a stand of mature Banyan trees through the creation of an artificial island represents ecological design strategies never experienced in Zhongshan or southern China.

Yu and Turenscape design strategies and approaches are noted in Table 5.2. Like the illustrative table for Living Water Park, Table 5.2 summarizes ZSP as a case study

Figure 5.32 Wedding photography along the park's railway path, photograph by author

Figure 5.33 Aerial photograph of ZSP, China's first re-purposed industrial site for a new public park, photograph by Turenscape

Table 5.2 Summary of case study two: ZSP's Hybrid Modern design language, graphic by author

Design approach	Design grammar	Design Vocabulary	Materiality
Spirit of Place [*genius loci*]	Preserves the site's industrial heritage and celebrates the memory of the shipyard factory workers	Industrial scenes with people depicting everyday life in the collective	Recycled industrial artifacts, re-imagined water towers celebrate the spirit of the workers and awaken a new society, human-scale bronze figurative sculptures of workers
Cultural: Local identity, nation-building	Symbolism	Rectilinear geometry and linear paths symbolize Mao's industrial age and the collective,	Railway path design; Grid paving pattern with steel bands; green rooms mimic dimensions of dormitories Preserve mature trees that define riverfront and shipyard boundary
	Healing memories of the Cultural Revolution	Red Box, a place for contemplation, is isolated in the park's landscape and open to the sky, bifurcated path	Red painted steel enclosure without a roof with interior reflecting pools and stone paths Unrestricted access to lawn areas provides newfound sense of individual freedom and post-Mao society
Didactic	Learning mid-20th century history of China, national memory of Mao's industrial era and political event; and local history of ship-building industry and factory collective	Living museum of everyday life at a ship-building communal factory Interpretative signage with explanations and poetry about political history	Recycled industrial artifacts & re-purposed building structures; landed boat as play area; bronze figurative sculptures posed in everyday work scenes
	Learning dynamics of the lake's fluctuating water levels	Pedestrian lakefront access and observation areas	Lakefront stepped path system of piers with native wetland plantings

(Continued)

Table 5.2 Summary of case study two: ZSP's Hybrid Modern design language, graphic by author (*Continued*)

Design approach	Design grammar	Design Vocabulary	Materiality
Ecological	Retain existing lakefront natural edge and reveal local natural habitat, reveal the lake's daily tidal changes, retain and re-locate existing trees, preserve mature stand of heritage trees	Waterfront access at lake and river A new island preserves heritage trees and accommodates flood control for the river Restore natural ecology	Pedestrian boardwalk system of stepped piers creates access to lake Stone used for embankments and walls for the ecological island Pedestrian access to naturalized lakefront edge, woodland and wetland plants
Chinese Picturesque	Scene-making	Waterscape	Lakefront pavilions to look from, look through, and look at from a distance
	Scene-manipulation	Borrowing scenes Framing scenes	Island in lake contains a pavilion to walk through, look from, and look at from a distance; Red Box creates enclosure and doorways frame views to look out and look in
	Yin-Yang	Passive/active zones	Spatial organization
Beaux-arts/ western design	Axial geometry as visual ordering principles	Axial paths as visual corridors and for organizing pedestrian mobility through the park's path system hierarchy	Machinery and industrial objects are used as visual focal points that terminate the visual axis at each path corridor Red Box as *folie*

for conceptualizing the hybrid modern design language. Furthermore, Yu's design contributed to China's emergent urban aesthetic of that time.

Many recycled and re-used industrial elements are obvious to the casual park visitor. However, it is noteworthy that the provenance of many of these artifacts would not be transparent to the general public. The general public might have difficulty grasping the concept of the recycled, re-purposed and reconstructed industrial artifacts. The use of these items clearly was designed with a wider audience in mind – governmental officials and the international design community. Additionally, Turenscape purposefully went against the grain of the modernist "tabula rasa" approach of clearing the site, erasing its industrial memory and Mao's heritage, and replacing it with an ornamental park without deep meaning.

The emphasis on re-use in ZSP stands in sharp contrast to some other notable legacy parks such as *Parc de la Villette* in Paris, one of Francois Mitterrand's *Grands*

Projets. The early 1980s international design competition sought a new vision for a 21st century public park in Paris on a former abattoir site. The winning entry by Bernard Tschumi was a self-conscious effort to create the first postmodern park, antithetical to notions of park design that took their cues from nature. His design proposal essentially erased the site's memory, or "spirit of place", and deployed a design strategy that incorporated a system of points, lines and surfaces: a type of collage comprised of different design layers and a series of "themed" gardens. Considered avant-garde, Tschumi's premise was oriented to a broader notion of creating a cultural experience for exploration, discovery and interaction (Baljon 1992). Due to its lack of sensitivity to the site's natural history or lack of acknowledgment to natural resources or the potential for ecology, the idea of Tschumi's design of *Parc de la Villette* remains contested among landscape historians.

Although it is largely unlike Parc La Villette in Paris, ZSP is a similarly "layered" design or "collage" of ideas. It combines several layers of information that address the site's socio-cultural history (industrial ship-building, communist work unit, Mao's era of extremes and contradictions), and shifting needs in the Deng New Era. Ecological design functions as another layer, especially the creation of an artificial island to address river flooding, and more significantly the preservation of mature stand of trees as a natural habitat and historic site edge.

The layers of meaning and symbolism for some of the design elements were intended to provoke visitors. The Red Box, for example, harkens back to the societal rupture during Mao's Cultural Revolution. Aside from this steel enclosure's "shock of the red" in the park's field of bamboo and lawn, the reflecting pools within the Red Box are meant to symbolize cleansing the dark period of the Cultural Revolution. The bifurcated path symbolizes the interruption and turbulence of this ten-year period of arrested cultural development (Figure 5.34). The Red Box is reminiscent of "Scar painting, *shanghen huihua*" and "scar literature, *shanghen wenxue*", a genre from the early post-Mao period when China's intellectual literati confronted the societal rupture created by the Cultural Revolution with explorations into spiritual healing. The steel and glass structure or light beacon, an encased re-used water tower located on the ecological island, was designed as a symbol for both China's cultural awakening and the preservation of its recent past (Padua 2003).

Turenscape's design was clearly influenced by their research on international design precedents: Haag's Gas Works Park and Latz's Duisburg Nord Park. ZSP preserves the site's industrial heritage and natural resources, and created didactic experiences for visitors to learn about local and national history, and the site's dynamic water environment. Turenscape's move to adaptively re-use existing buildings combined with new buildings also reflected an international trend that had already made a substantial impact in China. The high-profile pedestrian-oriented "*Xintiandi*" (literal translation is new heaven and earth with accepted contemporary interpretation: New World), commercial "lifestyle" development in Shanghai was completed just before ZSP's completion. This signature commercial development combined the adaptive re-use of Shanghai's historic buildings in the French Concession district with new structures made to look as if these were part of the original historic fabric.

The use of bronze figurative sculpture as a main focal point in a park is a characteristic feature of park design and public plazas throughout the world, including China. Yuezhong Shipyard's everyday scenes were created through human-scale bronze figurative sculptures and industrial remnants including the foundry's smoke

Figure 5.34 Bifurcated path and reflecting pools symbolize spiritual healing, photograph by author

stack and one of the gantries. These scenes were based on oral histories recorded by Turenscape. However, Turenscape broke convention in the way these elements were typically used in China. In China's public parks or squares, figurative sculpture was traditionally depicted in a larger than life or monumental scale (local poet, musician, folk hero, emperor, etc.). ZSP's figurative sculpture of everyday scenes were intentionally designed to mimic the human scale.

ZSP also illustrates the historical conventions of European design. In the spatial form of the park, Yu uses a layer of axial design geometry to set up visual and pedestrian corridors that connect the main elements of the park. This system of axial sight lines and focal points borrows from the western Beaux Arts tradition. In other areas of the park's urban active zone, he creates a visual rhythm using repetition in his use of the grid module for paving and raised planting bed patterns that are reminiscent of works by Peter Walker, Dan Kiley and other modern landscape architects. For Yu, the grid and rectilinear forms symbolize Mao's industrial heritage and his attempts to modernize socialist China.

Other imported design elements include the use of clipped hedges or parterres to mimic spatial enclosure and dimensions of the shipyard's residential dormitory rooms. Yu's intention is part didactic and part celebratory – teach visitors about Mao's work unit through an immersive experience and celebrate the socialist lifestyle past. Parterres, or the shaping of plants to create sculptural forms, is a horticultural practice that has its roots in both Chinese and European garden traditions. The Chinese traditional practice, or *penjing*, similar to "bonsai", however deals with the shaping of individual plant species into sculptural or "poetic" forms, usually in containers. It is believed that the use of parterres was imported during China's colonial era in places like the French Concessions in Shanghai and Guangzhou and the Italian Concession in Tianjin where public gardens were designed for foreign consumption. Additionally, French horticultural practices may have made its way to China via Li Ju, a horticulture professor at Zhejiang University who studied horticultural engineering in France at the National Horticultural Institute in the early 1920s (Lin 2005).

Open lawn areas were used throughout ZSP's active urban zone where visitors could socialize and picnic regularly. Open pedestrian access to lawn areas was relatively new at the time. Chendgu's LWP and ZSP were the exceptions. Most parks during that time, and even in some locations today in China, restricted physical access to lawn areas and typically enclosed them by a fence. For Yu, open access to the lawns was a symbol for China's break from the past, and represented individual freedom. Creating interactive experiences for visitors to observe the lake's tidal fluctuations were also new to China at that time and symbolized post-socialist society.

The influence of Chinese Picturesque design language is also evident at ZSP. At first glance, the park appears to be a direct denial of that influence with its emphasis on Mao's era. Yet the Chinese Picturesque design grammar is woven into the design in subtle ways. The lake, mature vegetation, and the stripped sheds along the lakefront are organized to form scenes and frame waterscape scenes in a tradition central to the Chinese Picturesque. The stripped sheds act as waterfront pavilions. The pavilion on the artificial island incorporated wooden steering wheels (Figure 5.35) and is referential to boat pavilions found in traditional Chinese gardens. In addition, the ecological island containing the preserved stand of mature Ficus trees, while intended for flood control, is also referential to the Chinese Picturesque island where immortals reside. An archival review of Turenscape's design drawings depict the use of scenic and pedestrian system diagrams consisting of an interconnected system of the park's eight scenes visual corridors, and pedestrian movement and paths. The use of scenes in Chinese Picturesque gardens is also typical. Perhaps the only feature of the Chinese Picturesque tradition that was omitted is rock piling. No artificial mountains, rock-piling or rockery were incorporated into ZSP's design.

ZSP's hybrid modern form in some ways represents a collage of design approaches. Not one single design approach was deployed. In interviews and written works by Yu on his inaugural award-winning project, he emphasizes "landscape design is a process that visualizes the meanings of the site through preserving and modifying the existing forms, and if necessary, creating new forms" (Yu & Pang 2002, p. 36). His work also wanted to reveal the beauty of the "weeds" or native grasses and indigenous plants as a progressive way of thinking about nature and an alternative to cultivated roses, chrysanthemums and peonies.

Yu's collage of design approaches (cultural heritage, didactic, *genius loci*/spirit of place, ecological, Chinese Picturesque) for the hybrid modern ZSP was heavily weighted towards Mao's nation-building efforts and the societal transformation that was occurring at the time. Cultural awakening was also expressed in symbolic design gestures throughout the park. While it is the first park in China to commemorate Mao's collective work unit with international design gestures, it is still imbued

Figure 5.35 The island, containing a pavilion with recycled ship wheels, is reminiscent of the Chinese Picturesque and creates views of the lake and Yu's deconstructed boat sheds, photograph by author

with subtle gestures toward the Chinese Picturesque. Turenscape's collage approach and fusion of the international design and Chinese Picturesque design, memory and meaning, nation-building and local identity are indicative of the concept for hybrid modernity.

In China, Turenscape's design of ZSP reflects a late 20th century revolution in park design. Yu has since evolved into a major voice in the discipline of landscape architecture in China and internationally. In some ways, ZSP reflects Yu's personal experiences. He has childhood memories of the Red Guard and Mao's Cultural Revolution, and the fact that his family suffered because of their affiliation as a so-called "landlord" or member of the bourgeoisie – targets of Mao's 1966–1976 dark period.

Yu is genuinely committed to the profession of landscape architecture and sees himself as a prophet to heal China's environment. He uses the ancient practice of binding women's feet as a metaphor for the arrested development of landscape architecture as a modern professional practice in China. From Yu's perspective, traditional foot-binding represents the Chinese Picturesque genre that is not allowing innovative thinking or progress in the discipline of landscape architecture in China. In effect, his voice represents the radical or revolutionary perspective, versus an evolution of the Chinese Picturesque that many of China's senior educators and design professionals advocate for and practice.

However, hybrid modernity does not necessarily align with Yu's disdain for the Chinese Picturesque. Often, Turenscape deploys Chinese Picturesque design conventions, primarily as a way to communicate with the older generation. On the contrary, Yu might be uncomfortable being characterized as using design grammar from the Chinese Picturesque in his work. The fact that even Yu would feel those features are necessary in his work bears testimony to the degree to which Chinese Picturesque elements have become a means of marking a design as "Chinese". At the same time, Turenscape's design for ZSP introduced innovative design approaches never experienced in China up to that point when the park opened in 2001. Like his counterparts, China's avant-garde New Wave artists, Yu takes on the responsibility to teach society about ways to celebrate the recent past (former Mao era industrial communal factory) and re-purpose these spatial forms into a new park where visitors can learn about the site's environmental qualities (fluctuating daily tides, and novel way to deal with river widening and preserve a stand of heritage trees, reclaimed natural habitats and restored natural systems) and industrial heritage, as well as cultural awakening through contemplation of the Cultural Revolution (Red Box symbolism) and a new park experience that symbolizes freedom and the post-Mao era.

Case study three: Jinji Lake Landscape Master Plan, *Jinji Hu Jingguan Zongti Guihua*, Suzhou, Jiangsu Province

> "Our philosophy was based on 'one mirror, two reflections' and duality between symbolic traditions of Suzhou and modern life in China".
>
> Chiao, S. interview, 2003 March

The third case study, Jinji Lake Landscape Master Plan (JLLMP), originally by EDAW, Inc. (since merged with AECOM in 2009), is larger in scope and scale than the hybrid modern parks in Zhongshan, Guangdong, and Chengdu, Sichuan. This EDAW Legacy project is located in a 1990s expansion area or "new township district" known as Suzhou Industrial Park east of "Old Suzhou", Jiangsu Province, the accepted historic capital of Gardens of the Literati – private residential gardens designed and built for retired imperial government scholar-officials since the 11th century (Figure 5.36). This legacy EDAW project covers approximately 800 hectares (1977 acres) around a seven square-kilometer (2.7 square miles) freshwater lake in the China-Singapore Cooperation Zone, a national-level development pilot project or new town model. Design components of EDAW's award-winning Jinji Lake Landscape Master Plan (JLLMP) along the northern and eastern lake waterfronts were completed by 2002–2003.

Suzhou context

Suzhou, a secondary Chinese city, is located in the Jiangsu Province and is part of the Yangtze River delta region. This ancient canal city was established in the sixth century

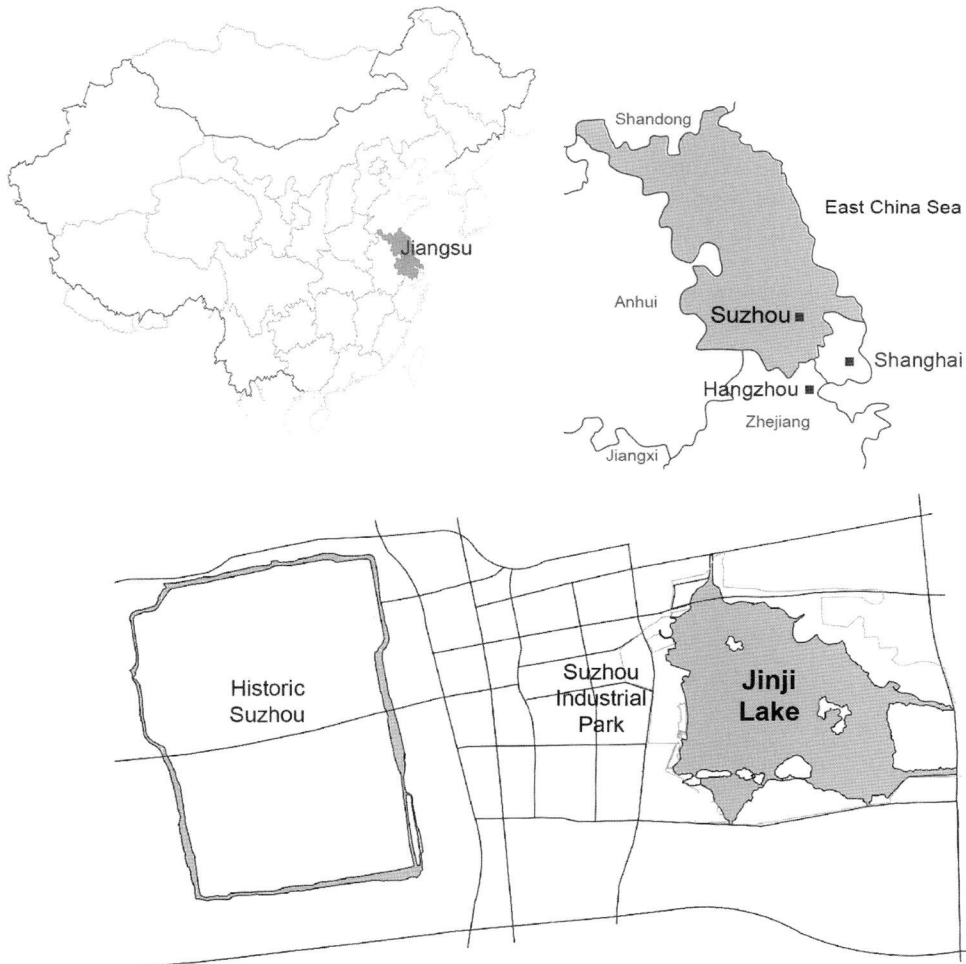

Figure 5.36 Context and location map for Jinji Lake Landscape Master Plan, graphic by X. Liu

BCE, and is located 112 kilometers (70 miles) west of Shanghai. Before the build-up of Yangtze River's silt and changes to the delta region over time, Suzhou had direct access to the East China Sea via the Wusong River, also known as Suzhou Creek (Xu 2000). Like Chengdu, it was a capital city during the Warring States or Three Kingdoms period. Later, during the imperial era, it emerged as one of the major administrative and commercial centers along the Grand Canal, China's historic major mode of transportation for moving water and goods from the water-rich south to cities and towns in the water-scarce north.

Throughout the centuries, Suzhou developed into a significant hub for agriculture, commerce related to both cotton and silk-based textiles, land ownership, literati culture and intellectual lifestyles. Given Suzhou's morphology was based on a system of canals and its status as a commercial and cultural hub where government scholar-officials retired, it hailed the reputation as "Venice of the East". A popular term inspired by traditional poetry noted, "*shang you tiantang, xia you Suhang*, in heaven there is paradise, on earth there is Su[zhou] and Hang[zhou]", and evoked the notion that both Suzhou and Hangzhou were considered "paradise on earth" (Xu 2000; Wang & Bramwell 2012). Suzhou's cosmological and urban form has been the subject of in-depth studies, especially as an exemplar of China's imperial city planning and nation-building over time.

Similar to major urban centers in southern China during the Taiping Rebellion (1850–1864), Suzhou declined dramatically. Much of the city fabric was destroyed, especially when it was occupied by Taiping leader Li Xiucheng; and much of Suzhou's population fled to Shanghai, a place unconquered by the Taiping. Eventually, during post-Taiping reconstruction, Suzhou's silk industry was able to compete in world markets when European and Middle Eastern silkworms were killed in an epidemic circa 1854 (Marme 2018). Suzhou's commercial and world export status would prove more significant as a result of the 1895 Treaty of Shimonoseki that ended the first Sino-Japanese War (1894–1895). At that point, Suzhou was established as a treaty port and foreigners took advantage of opportunities to establish factories in China's interior. Additionally, this unequal treaty caused the expansion of foreign concessions, especially Japan, in both Suzhou and Shanghai. When a railway between Shanghai and Suzhou was built in 1906, it reduced a twelve-hour steamboat ride to just over two hours by train. This shifted the balance toward Shanghai as the modern commercial center with Suzhou as a center for traditional culture and handicraft industries.

Like many urban locations after Mao's communist revolution and 1949 establishment of the People's Republic of China, Suzhou's population declined. As part of Mao's Great Leap Forward (1956–1960), or "grain and steel" nation-building policy and shift from the "consumer city" to the socialist "producer city", the population in and around Suzhou relocated to four new industrial towns within ten to twenty kilometers of historic Suzhou. Mao's Cultural Revolution caused further destruction to much of Suzhou's historic fabric including the loss of ancient city walls and the majority (around two-thirds) of traditional Chinese gardens. Suzhou's urban development, like many cities throughout China, was stagnant between 1960 and 1982. However, like many rural areas throughout China in the 1980s, local industries around Suzhou and the Yangtze River delta area experienced economic success through the Town and Village Enterprises program, the first wave of Deng's reforms mentioned in an earlier chapter.

Soon after Deng's reforms were launched, Suzhou, along with Chengdu, Sichuan and Hangzhou, Zhejiang (location of the fourth case study park) were listed in the first batch of twenty-four Chinese cities designated as National Historical and Cultural Cities in 1982 by the Ministry of Construction (re-structured as the Ministry of Housing and Urban-Rural Development in 2008). The establishment of this list was an expansion of the 1982 inaugural launch of China's Law on the Protection of Cultural Relics and the Ministry of Construction and called for prescribed guidelines. In this light, Suzhou's city government officials moved forward to develop the 1982–2000 Suzhou City Master Plan with an overall vision of "one body with two wings" (Figure 5.37). Suzhou's historic walled city would serve as the body flanked by wings or expansion areas on its east and west sides. The west expansion area was referred

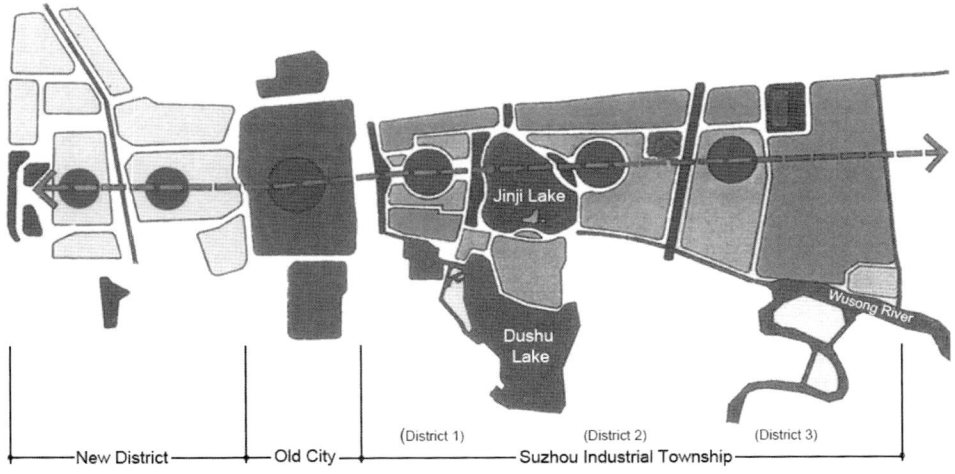

Figure 5.37 Suzhou 1982–2000 City Plan, based on a map provided by Suzhou Industrial Park
      Administrative Committee, published with permission

to as Suzhou New and High Tech District (SND), and the area east of Old Suzhou as
Suzhou Industrial Park (SIP).

A portion of the SIP eastern expansion area evolved into a designated national-level
project and formal cooperative effort between Singapore and China. Its roots lay with
Deng's 1978 diplomatic visit to ASEAN (Association of Southeast Asian Nations)
countries (Thailand, Malaysia and Singapore) and later 1992 "Southern Tour" of
China including visits to Shanghai's Pudong District, Shenzhen and Zhuhai Special
Economic Zones (SEZs), Zhonghsan and other key southern cities. Deng's 1978
ASEAN itinerary included visits to Singapore's Housing and Development Board and
Jurong Town Corporation where he learned about the successes of "foreign direct
investment" utilized in the development of the Jurong industrial township, originally
a marshland (Lee 1981). Deng also met with top officials from Singapore's banking
and insurance industry.

Deng called for increasing the pace of reform and opening-up during his later 1992
"Southern Tour" after seeing firsthand the successes of his reforms in key southern
Chinese cities (Lee 1981; Zhao 1993). Media coverage of Deng's speeches and pro-
nouncements of his nation-building efforts during his 1992 tour also noted "Singapore
as a capitalist version of the communist dream" (Perry & Yeoh 2000). Deng's state-
ment resonated with Singapore leaders and eventually led to the selection of Suzhou
for a joint national development effort, modeled on the Jurong modern industrial
township. In fact, the subsequent 1994 Sino-Singapore Agreement on the Joint
Development of Suzhou Industrial Park (SIP) was similar to Mao's 1950 Sino-Soviet
agreement in that it involved both financial aid and a so-called "software transfer" – a
transnational policy that enabled the importation of Singapore's public administration
and economic management practices, especially the industrial development model, to
China via Singaporean experts (Perry & Yeoh 2000).

The administration of the Sino-Singapore SIP cooperation agreement required the establishment of two key bodies: 1) the China-Singapore Suzhou Industrial Park Development Company (CSSD) made up of representatives from Singapore and China who were charged with seeking out foreign investors and land development opportunities for the former agricultural area and marshland, and supported by 65% Singapore government financing and 35% Chinese government financing; and 2) Suzhou Industrial Park Administrative Committee (SIPAC) with oversight for the planning and implementation of projects in Suzhou Industrial Park, east of historic Suzhou. Over the course of a year or so, Chinese officials were trained by Singaporean experts, either in Singapore or Suzhou, on policies and practices that included new township planning, land development and finance, infrastructure engineering and public administration. Singapore experts taught Chinese officials the foreign-investment model utilized in their Jurong industrial township and other imported British ideas on new town planning precedents.

Since its inception, SIP was referred to as a new industrial township, a type of land development, modeled on Singapore's Jurong Industrial Estate. Singapore's model was established in the 1960s and transformed marshy swamp areas into a key manufacturing area funded by foreign direct investment; and it created thousands of new jobs and new housing (Huat 2011). Singapore, a former British colony, adopted much of the UK's technical urban planning nomenclature for planned communities which included "new town", "township" and "housing estate". The 1994 Sino-Singapore SIP joint development sought to create jobs for high technology and related industrial and manufacturing sectors for foreign expatriates and local Chinese, as well as new housing for these employees. Additionally, SIP's goal was to create a balanced new town with hi-tech employment opportunities and housing for a population of 600,000, as well as a new central business district for historic Suzhou (Lin 2015). In urban planning nomenclature, SIP was essentially a "new town" and at the time of EDAW's 1998–99 commission for the Jinji Lake Landscape Master Plan, planned communities in the US were referred to as "new communities" or lifestyle communities.

One of SIPAC's early efforts involved the cooperation between Singapore's Urban Redevelopment Authority (URA) and Suzhou's Planning Bureau to formulate the 1994 SIP Master Plan based on Suzhou's "one body two wings" vision. It also established land use, zoning and development densities for eight districts in the SIP Singapore-China cooperation zone and slated for development in three phases. The total area of SIP covers 288 square kilometers (112 square miles). Early implementation phases of SIPAC's efforts involved infrastructure and road construction. This included dredging and draining the water in Jinji Lake to excavate needed soil from the lake's bottom to re-shape and increase the height of the surrounding landscape for flood control purposes.

Project inception

Joseph Brown, landscape architect and former CEO and President of EDAW, Inc., (fully merged with AECOM in 2009), and the Pei Group, New York-based architects, were part of a consulting team engaged by Suzhou Mayor Zhang Xinsheng in 1996. Their commission involved the formulation of design guidelines for the redevelopment of Pinjiang, one of ancient Suzhou's historic canal districts (Prentice 1998). This experience led to SIPAC's subsequent correspondence with EDAW and eventual design commission that commenced around 1998–1999 (Padua 2004a).

Sean Chiao, AECOM Asia Pacific President and former director of EDAW's Hong Kong office, noted SIPAC initially called for the landscape design of a 200-meter wide band along the perimeter of Jinji Lake. As part of their due diligence, EDAW's team analyzed the previously approved 1994 SIP Master Plan document by URA and Suzhou Planning Bureau that included land use, development density and zoning and district level guidelines for the designated eight districts around Jinji Lake's perimeter. Recognizing that their design commission was focused on a national-level development project and model for other Chinese cities, Chiao convinced SIPAC to expand the scope of their contract. EDAW's JLLMP created new guidelines for a system of public open spaces that integrated and linked the eight districts internally and externally with Jinji Lake's waterfront. Given SIPAC's initial construction phases were focused on building critical infrastructure, like public utilities and road systems, EDAW pointed out that the system of new public open spaces around Jinji Lake were also significant infrastructure components needed for SIP's successful development.

Extending further the knowledge transfer from Singapore to China, SIPAC called for a formal joint venture between the American-based firm, EDAW, and the city of Suzhou's Garden Design Bureau (Padua 2004a). As noted in the LWP case study narrative, foreign design firms typically didn't participate in the construction process. In the case of the design and implementation of JLLMP, the formal joint venture allowed EDAW's direct engagement during construction and helped to ensure the quality of construction would represent their design intentions. At the same time, Suzhou's local design team was introduced to western standards of professional landscape architecture practice and corporate culture espoused by EDAW, one of the largest planning and design firms with numerous branch offices around the world at that time.

Project realization

The "Legacy EDAW" (AECOM's reference to EDAW projects completed prior to 2009) award-winning project, Jinji Lake Landscape Master Plan, was largely an urban design scheme driven by a system of public open spaces geared towards the new community of foreign expatriates and local professionals. EDAW's scope of services entailed major revisions to the detailed plan for over 800 hectares (1977 acres) previously approved by SIPAC. EDAW's Landscape Master Plan contract involved evaluation of SIP's land use and development density, modifications to the lakefront edge, re-design of overall circulation (pedestrians, bicycles, motorized vehicles), and new definitions of public open spaces and built form edges for the eight districts surrounding Jinji Lake (Figure 5.38).

The size and scale of the JLLMP called for a team of design and planning professionals drawn from EDAW's San Francisco and Hong Kong offices. Initial stages of their work involved research on successful waterfronts. The team's design philosophy was culture-based and acknowledged the local water town heritage, Suzhou's traditional Chinese Picturesque garden heritage and the emergent communities or "New

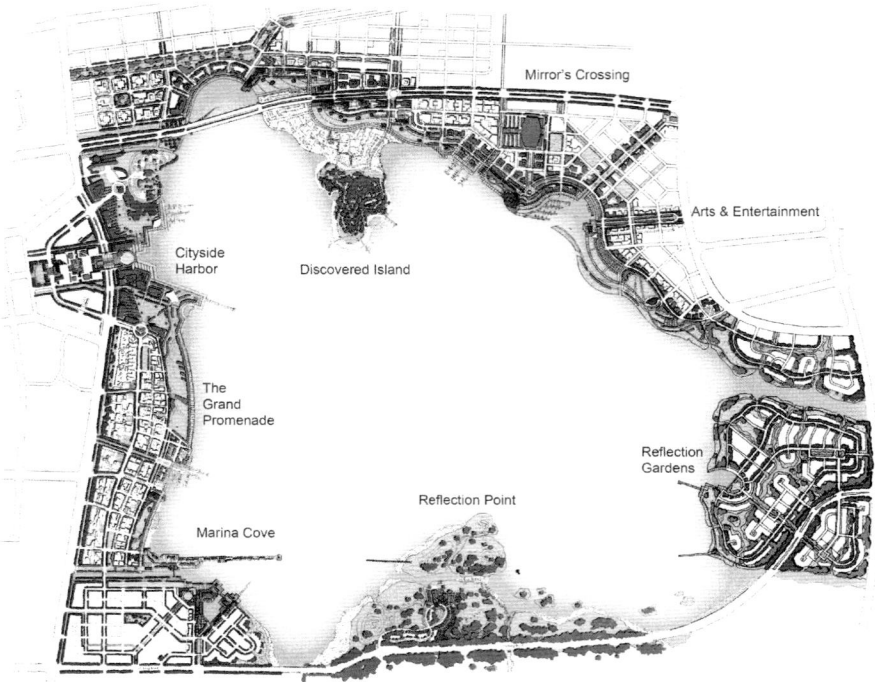

Figure 5.38 Jinji Lake Landscape Master Plan, Legacy EDAW project, fully merged with AECOM in 2009, graphic by AECOM

Suzhou" in SIP's Sino-Singapore Cooperation Zone. EDAW's design team's ensuing design philosophy for Jinji Lake was described as "one mirror, two reflections" – a duality between symbolic traditions in Suzhou and post-Mao contemporary China lifestyles (Padua 2004a). One mirror, SIP's image as an international place for living and working, reflects: 1) "New China's" contemporary lifestyle; and 2) local heritage consisting of the area's water towns and traditional Chinese gardens. EDAW's work was informed by research on local heritage and successful waterfront developments and livable communities around the world.

Presentations of EDAW's proposal for Jinji Lake's public open spaces and designed outdoor environments were televised regionally and SIPAC's review and approval were non-controversial. As a national development project and formal joint venture with Singapore geared for foreign direct investment, SIPAC acknowledged the importance and impact of Jinji Lake's image and ways that designed landscapes could foster a supportive FDI climate. SIPAC also understood the leisure and consumer needs of nearby Shanghai residents and the emerging private car-oriented society. Furthermore, at that point, the regular train ride between Shanghai and Suzhou was just over one hour. Currently, high-speed trains between Shanghai Hongqiao airport and Suzhou take less than thirty minutes.

Working on a so-called "greenfield" site and agricultural land, a location with no previous history of urban development in itself would be daunting for most landscape architects and planners today. However, as a national experiment for a model Chinese city, the design team realized the significance of their work. While restricted somewhat by the previously approved SIP 1994 Master Plan developed by Singapore's URA and Suzhou City Planning Bureau, EDAW designers took care to formulate guidelines that modified development densities to support the most active public open spaces. Other guiding design principles involved retaining the lake's natural shorelines and restructuring land uses in the eight neighborhoods to optimize green open space linkages and create a system of human-scale public open spaces along the lakefront. EDAW's team also introduced ecological design principles by incorporating best practices for stormwater management and minimized surface water run-off into Jinji Lake.

EDAW's initial phase of JLLMP was completed by 2002–2003. SIPAC considered these public open spaces critical components of Jinji Lake's overall infrastructure. The system of interconnected and accessible public open spaces along Jinji Lake's waterfront and within the eight districts were perceived by SIPAC as assets that improved land values for adjacent properties slated for foreign investment (Lin 2015). In EDAW's realization of JLLMP, broad design strokes were taken. EDAW spatially organized the project into two zones (active and passive) around an axis that traversed Jinji Lake from the northeast to the southwest. The upper half was "active" and designed to accommodate intensive development that emphasized commercial and recreational uses including arts and entertainment, commercial districts, and an area of high-density housing with a marina. The lower zone was characterized "passive" with proposed waterfront housing, a variety of other housing types, neighborhood amenities and a regional wetland park.

The design team was both challenged and inspired by the Chinese Picturesque genre depicted in the nearby traditional scholar gardens of historic Suzhou. Completed designed landscapes at Cityside Harbor, also Suzhou's new central business district, re-interpreted the Chinese Picturesque traditional gardens. In the adjacent Grand Promenade district, EDAW's team based their ideas on western waterfront precedents. Critical to the design of these public open spaces in both districts was the full deployment of Chinese Picturesque scenery manipulation, particularly borrowing the view of Jinji Lake's vast waterscape scenery, a tremendous backdrop and landscape feature.

While the original design intention sought to maintain the lake's natural edge, the western waterfront along the City Harbor and Grand Promenade districts needed major modifications. Some areas were filled in to construct key designed landscapes such as large open lawn areas, hilly topography for the Camphor forest garden, large public plaza and waterfront promenade. At the same time, other sections of the lake's edge were re-shaped and cut to form basins for the proposed marina. Other areas were modified to establish soft lake edges with "naturalized" landscapes, wetlands and new wildlife habitats. EDAW derived overall identities for each of the eight districts and their work was realized in Cityside Harbor, Grand Promenade and part of the Arts and Entertainment district.

The identity for Cityside Harbor's waterfront was largely urban in its character. It was designed to accommodate office workers from the adjacent commercial district and as historic Suzhou's new central business district. EDAW's design re-interpreted

Figure 5.39 Pavilion in Camphor Forest, photograph by author

concepts related to the region's water or canal town heritage and elements of the Chinese Picturesque. Their work created a contemporary design narrative for under-standing these local traditions. At its completion in 2002–2003, this district included two large public plazas with vast fields of hardscape paving, a forest garden comprised of 1000 camphor (*Cinnamomum camphora*) trees on an artificially constructed hill, lakefront rock garden with wading pools, pedestrian path system with areas for sitting, covered walkway and a canal that linked Jinji Lake to historic Suzhou. The articulated hard edges of Jinji Lake's waterfront in Cityside Harbor included an amphitheater and wooden pier, and segments that stepped down to address tidal changes along with physical access to the water for visitors.

EDAW's Camphor Forest garden in Cityside Harbor creates the appearance of an established naturalized landscape for use by SIP's workforce, residents and tour-ists. One thousand mature camphor trees were installed atop an artificially made hill next to the waterfront rock garden. Within the forest, a pedestrian path system made up of stepping stones in fields of grass traverse the undulating topography to hill-top pavilions (Figure 5.39). These wood-framed and glass-covered pavilions are contemporary and minimalist in style and perform functions similar to the "*ting*", small structures typically found in Suzhou's traditional private scholar gardens. Like the traditional "*ting*", pavilions in the Camphor Forest are sheltered resting places that offer views of the surrounding landscape with Jinji Lake's waterscape as a sce-nic backdrop. EDAW's client was enamored of the Camphor Forest and requested the inclusion of another forest garden across the lake in the proposed Arts and Entertainment district. To contrast the evergreen (non-deciduous) Camphor Forest

Figure 5.40 Covered walkway links Camphor Forest to Jinji Lake waterfront, photograph by author

Figure 5.41 Wading pools accessible via pedestrian piers connected to covered walkway, photograph by AECOM

Figure 5.42 Waterway links New Suzhou to historic Suzhou, photograph by AECOM

garden, EDAW's design involved the installation of the Maple Forest garden (1000 trees), also on an artificial hill, and consisted of mature trees with leaves that turned red in the fall season.

A covered walkway is purposefully designed as a reference to the *"lang"*, another Chinese Picturesque element found in Suzhou's traditional gardens, which connects the Camphor Forest to the rock garden and wading pools at the lakefront (Figures 5.40 and 5.41). The *lang* is a device used to control the view in an ordered "slow" experience where scenery or series of scenes are revealed while walking. It was also used to create the illusion of space and is referential to the Chinese Picturesque garden. Former EDAW employees and principal designers David Jung (AECOM Vice President) and Frank Chow (Principal of Hong Kong-based FRC studio) deployed the *lang* and related series of perpendicular paths that lead into the lake's wading pools to do the opposite – create a human-scale experience adjacent to Jinji Lake's vast waterscape and visual horizon line. Large boulders defined the lake's original natural edge and created places for visitors to access wading areas in Jinji Lake and enjoy the panoramic waterscape views. The system of linear pedestrian piers perpendicular to the covered corridor path create access to the wading pools and rock garden in Jinji Lake (Figure 5.41), and also create a visual foreground and human-scale experience, another form of Chinese Picturesque scenery manipulation.

Strolling back to the commercial district in Cityside Harbor, the scenes shift from Jinji Lake's waterscape promenade path to another urban-oriented waterfront pedestrian experience. This urban scale promenade path fronts a canal that links Jinji Lake to the canal system of old Suzhou. In the visitors' field of view, the skyline of Cityside Harbor serves as borrowed scenery (Figure 5.42). The canal-front curvilinear path is designed with contrasting light and dark stone bands, which are symbolic references to the dark roof tiles and white walls of traditional low-rise buildings found in local water towns. In effect, EDAW's design incorporated a waterway that links SIP or "New Suzhou" to historic Suzhou. This waterfront path also connects to the series of public open spaces along Grand Promenade's waterfront.

The Grand Promenade lakefront design and system of public open spaces is largely referential to western design precedents. The only Chinese Picturesque aspect of the new public open spaces was the use of borrowed scenery – Jinji Lake, a panoramic waterscape view. To inform the broad design strokes in this district, EDAW's design team studied the waterfronts of Sydney in Australia, Singapore, San Francisco, California and other international precedents. Their final design was comprised of a two-tiered lakefront promenade path (a wide waterfront path, and the other adjacent tree-lined path with seating nested in vegetation oriented to the lake), and large unrestricted open lawn areas for informal recreation, leisure and social activities. The Grand Promenade's large open lawn areas are reminiscent of San Francisco's Marina Green sans the parking lot and Golden Gate Bridge. During field research on spring weekends in 2003, 2007 and 2012, the author observed various activities in the lawn areas including groups of children flying kites, picnicking and enjoying themselves, wedding parties and informal gatherings of families and young couples. Like Living Water Park and Zhonghsan Shipyard Park, access to the public lawn areas in the Grand Promenade district was open and unrestricted, unlike fenced and restricted lawns typical for public parks that also required entry fees at that time. The southern portion of the Grand Promenade lakefront transitions to a wetland park and incorporates native plants for wildlife habitats and stormwater management.

Larger in scale when compared to Chengdu's Living Water Park and Zhongshan Shipyard Park, EDAW's completed lakefront works were successful at creating a series of outdoor designed environments that visitors were able to enjoy. Contributing to its success was the quality of construction for a project of this scale. As noted earlier, foreign design firms were rarely engaged in a project's construction phase. However, the joint development with Suzhou's Garden Design Bureau enabled EDAW to participate regularly during the construction stage of the project when they learned about the limitations to materials and construction. For example, standards for wood or timber construction at that time were considered problematic. Jinji Lake's completed system of public plazas, promenades and parks in 2002–3 was heavily used by Shanghai residents during weekends and design elements made of wood were in constant need of repair.

Hybrid modern synthesis

EDAW's joint effort with Suzhou's Garden Design Bureau and the implementation of JLLMP's key components illustrate critical dimensions of hybrid modernity. This infrastructure project was foundational for China's new post-socialist image and intertwined with the goals of SIPAC's Sino-Singapore cooperation agreement. The central government's goal was to develop a new community for international consumption that balanced the high technology sector and sophisticated employment base with high-quality housing, as well as a new development prototype for China. In effect, the JLLMP was a key part of late 20th century nation-building in China.

EDAW's broad design strokes represented in JLLMP's system of open spaces were inspired by Suzhou's water town heritage and the fusion of the Chinese Picturesque garden tradition and international design influences. At the same time, EDAW deployed ecological design strategies that incorporated best management practices for stormwater management and the creation of wildlife habitats. The quality of construction also demonstrated the area's artisan heritage and high-quality craftsmanship. Like Damon's LWP and Turenscape's ZSP, EDAW's work at Jinji Lake emerged as a major destination and study model for China's CPC leaders and officials from various levels of government.

Along Jinji Lake's waterfront, the heart of the Sino-Singapore cooperation zone, the 2002–2003 completed works highlight the concept of hybrid modernity. Jinji Lake itself serves as a tremendous backdrop and Chinese Picturesque scene around which all of the completed works are organized. SIP employees, residents and tourists enjoy panoramic views of the lake and a vast waterscape scene from stationary points or while walking, bicycling or driving. They can enjoy open views while playing in Jinji Lake. They can also enjoy the feeling of the expanded open space that the lake's waterscape creates while socializing or flying kites in the large grassy waterfront meadows in the Grand Promenade District. In addition, they can experience a distant view while sitting in a pavilion in either the Camphor or Maple Forest Gardens. The variety of waterscape scenes celebrates Suzhou's water town heritage and reimagines the Chinese Picturesque scholar garden with contemporary design language. EDAW's work reinforced the area's new post-Socialist identity as an international livable community and a high-quality location to work and live. JLLMP's hybrid modern design, a major late 20th and early 21st century landmark project, fuses international design influences, the Suzhou scholar garden tradition and local water town heritage while reimagining local identity. It also symbolizes post-Mao and post-socialist nation-building efforts. At the time, it emerged as a weekend destination for Shanghai residents seeking new recreational and leisure experiences from urban life.

EDAW's 2002–2003 completed works for Jinji Lake Landscape Master Plan were several times the scale and size of both Living Water Park and Zhongshan Shipyard Park. Its hybrid modern forms were represented in broad design strokes that combined international design precedents for waterfronts, ecological design and the Chinese Picturesque. EDAW's work represented a hybrid modern narrative that also took into account local identity – a renewed image for Suzhou as a whole, and a reimagined identity for SIP. As a key infrastructure component for the Sino-Singapore SIP cooperation area and new community national prototype, nation-building was also a critical dimension for informing ways that JLLMP illustrated hybrid modernity. Table 5.3 summarizes the vocabulary, grammar and JLLMP's hybrid modern design language. As part of China's vanguard in the discipline of landscape architecture, JLLMP effectively represents the fusion of international design influences, the Chinese Picturesque and Suzhou's scholar garden tradition, local water town heritage, a reimagined local identity and nation-building – critical dimensions for defining post-Mao hybrid modernity.

Table 5.3 Summary of case study three: JLLMP's Hybrid Modern design language

Design approach	Design grammar	Design Vocabulary	Materiality
Cultural	Celebrate Suzhou's water town heritage	Incorporates water canal that links Jinji Lake to Old Suzhou along with the overall pedestrian path system.	Constructed edges step down to the canal and allow physical access
	Celebrate Suzhou's Traditional garden heritage	Pavilions in Camphor Forest Covered Walkway (*lang*) Waterfront rock garden	Covered corridor path links urban area, Camphor Forest (contains pavilions) to lakefront, wading pools and rock garden
	References to vernacular architecture	Paving design alternates contrasting bands for major paths	Represents the dark roof tiles and light walls of traditional low-rise buildings
Chinese Picturesque	Scene-making	Jinji Lake	Primary Waterscape scene and design focus, borrowed scenery
	Scene-manipulation	Camphor Forest, Maple Forest	Scenic Experience Pavilions (*Tings*) on hilltops as stationary places to enjoy mountain view and framed views of the lake Covered corridor path (lang), reveals rock water garden and frames views of the lake
	Rock piling	Artificial mountains	Green (Camphor) hill Red (maple) hill
Genius loci/ Spirit of Place	Symbolic references to site's physical history	Reconstruct the lake's original edge	Boulders and rocks define the original edge in wading pools and rock water garden Native wetland plants along segments of lakefront

(Continued)

Table 5.3 Summary of case study three: JLLMP's Hybrid Modern design language (*Continued*)

Design approach	Design grammar	Design Vocabulary	Materiality
Western	International waterfront design precedents	Waterfront promenade	Grand Promenade district: Two-tiered lakefront promenade path system
		Large grassy meadows	Large expanses of physically accessible open lawn areas
		Recreational water play	Wading pools at the lakefront
Ecological Design	Stormwater management	Wetlands design	Native wetland plants alongsegments of the lakefront
	Reclaim site's natural ecology and create new wildlife habitat	Green ecological park in southern district	Native plants
Nation-building	Landscape Master Plan represents prototype for a national development model for a new community for a professional workforce in high technology with housing & amenities for foreigners and emerging local professionals	Establishes system of open spaces that integrates eight districts with recreational open space system surrounding Jinji Lake	Reimagines Chinese Picturesque tradition of nearby Suzhou gardens and celebrates the area's water town heritage

Case study four: West Lake Southern Scenic Area Master Plan, *Xihu Huanhu Nanxian Jingqu Zongti*, Hangzhou, Zhejiang Province

"Our inspiration came from paintings of West Lake created during the Qing Dynasty".

W. Zhou (2007, pers. comm., 7 May)

The fourth case study, West Lake Lakeside Southern Scenic Area Master Plan (SSAMP), *Xi Hu Huanhu Nanxian Jingqu Zongti*, also known as the West Lake Southern Side Renovation Project, *Xi Hu Nanbian Chongxinzhuangxiu Jijian*, is located in Hangzhou, Zhejiang Province (Figure 5.43). The Hangzhou Landscape Architecture Design Institute (HDI), affiliated with Hangzhou's municipal government and comprised of locally educated professionals, spearheaded the detailed design of West Lake's Southern Scenic Area (Padua 2014). The SSAMP covers nearly 300 hectares (740 acres) and is located along West Lake's fourth side or "urban edge". As the fourth exemplar of hybrid modernity, this case study illustrates the inter-relationships among the key elements: the fusion of the Chinese Picturesque and international design influences, local identity and post-socialist nation-building.

Figure 5.43 Context and location map for West Lake's Southern Scenic Area, graphic by X. Liu

Similar to EDAW's work in SIP, the SSAMP grew out of a broader effort to renew Hangzhou's new post-Mao identity (Padua 2014). The significance of West Lake with its cultural and natural resources was vital to the city's grand exercise to reimagine its image. This hybrid modern project, circa 1999–2002, was instrumental for this effort, especially given that West Lake is of national-level significance as an international icon in Chinese literature and the arts for hundreds of years. HDI's hybrid modern design incorporates ideas and approaches that reflect international trends and contemporary interpretations of West Lake's folklore, mythology, urban heritage and cultural landscapes, especially the "*shan-shui*", mountain-water, landscape literati tradition. Historic preservation strategies for West Lake's cultural relics were also deployed, as well as ecological design strategies for restoring and conserving natural resources.

West Lake, Hangzhou context

Hangzhou is the capital of Zhejiang Province in southern China and is located in Hangzhou Bay, a major water body in the Yangtze River Delta. It lies south of the Yangtze River and like Suzhou is located in China's "south", an area historically known as the more sophisticated and cultured part of the nation. The mega-region north of the Huai River, where rice cultivation is considered problematic and wheat growing thrives, is generally considered less sophisticated and more agrarian (X. Sun 2008, pers. comm., 18 June). The general scholarly view is that the Huai River serves as the natural north-south boundary between the wheat-growing region of the rural north and the rice-growing region of the south where China's cultivated population thrived. Like Chengdu, Zhongshan and Suzhou, Hangzhou does not have the status of Shanghai, Beijing or Guangzhou, and is considered a secondary city in this book.

Hangzhou's long urban history reflects China's cultural, historical and political development. It was an ancient capital city for the Wu and Yue kingdoms in the Spring and Autumn period and a major administrative center in the Qin dynasty. It was the southern terminus for China's Grand Canal, *Da Yunhe*, a vast infrastructure project that commenced in the Sui dynasty with modifications made over hundreds of years and several dynasties (Wang 1999). The Grand Canal connected five rivers (Hai, Yellow, Huai, Yangtze and Qiantang) and was China's major mode of transporting agricultural products and goods from the water-rich south to the water-scarce north (Wright 1977). Hangzhou's commercial and cultural reputation expanded to international prominence when it became China's temporary capital during the Song dynasty when it was known as Li'nan (Clunas 1997).

Along with Suzhou, Hangzhou's idealized reputation was based on the popular saying mentioned earlier, "in heaven there is paradise, on earth there is Su[zhou] and Hang[zhou], *Shang you tiantang, xia you Suhang*", or simply noted as "paradise on earth"; and its identity was linked to West Lake, a major center of elite literati culture and popular religious activity (Wang & Bramwell 2012). In the mountains surrounding West Lake, numerous Buddhist and Taoist temples were built during the Tang and Song dynasties, establishing the area as a major religious pilgrimage destination (Chen 2003; Wang 1999; Wright 1977). The imperial capital grew to a population of 1 million and as the city's economy thrived, the literati arts also flourished. Subsequently, China's first visual arts institution, the Song Imperial Painting Academy, was founded in Hangzhou (Clunas 1997; Fairbank & Goldman 2006). Marco Polo, the Venetian explorer, wrote about the splendor of Hangzhou after he visited the city in the later Yuan dynasty circa 1290 (Chen 2003; Ebrey 2008; Johnston 1991).

West Lake, *Xi Hu*, and its link to Hangzhou's identity, is a significant part of China's landscape legacy – rich with ancient folklore and mythology (Keswick 1978; Chen 2003; Johnston 1991). One creationist myth notes that a pearl or white jade from heaven was carved and polished by a dragon and phoenix, and as they fought for ownership, the pearl fell to earth and transformed into West Lake; while the fallen dragon and phoenix transformed into the lake's surrounding mountains (Wang & Bramwell 2012). In terms of its natural form, the terrain was originally a small bay on the Qiantang River. It was transformed into a lake during the Sui dynasty when dikes were installed; and the lake functioned as the city's water and food source for Hangzhou, as well as for agriculture (Wright 1977; Chen 2003). West Lake's classical lay-out of the two tree-lined causeways named Bai and Su, and three artificial islands, were a result of repeated dredging of the lake over thousands of years. The combination of the lake's waterscape with "cloud-capped" mountains and hills surrounding the lake's three sides with the city on the fourth is a spatial form that has been unchanged for hundreds of years (Padua 2014).

The combination of mountains and lake created an idyllic subject for the Chinese literati and was the source of Hangzhou's reputation as China's cultural center for the so-called *shan-shui* (mountain-water) landscape tradition (Steinhardt 1999). Nearby Suzhou's reputation as China's garden capital was a result of private residential garden-making for and by retired imperial government scholar-officials; and Yangzhou evolved as a garden center as a result of the wealthy merchant class. The West Lake aesthetic was the focus for Tang and Song dynasty poets, musicians and artists when the "ten poetically named scenes" or ten classical scenes were realized and immortalized by poets and painters – serving as an inspiration for hundreds of years. When Buddhism influenced China's literati culture and the temples in West Lake's "cloud-capped" mountains emerged for cultural and religious consumption, West Lake's landscape aesthetic emerged as an inspiration and cultural phenomenon among Japanese and Korean artists and garden designers.

West Lake's landscape aesthetic, cultural tourism and Hangzhou's development declined in the Yuan dynasty when the new Mongolian emperor established Beijing as the permanent imperial capital city. In the following Ming dynasty when western influences in the arts and sciences from the Jesuit missions made their way to the imperial courts, China's literati gaze shifted away from West Lake. Emperors in the later Qing dynasty, who perceived themselves as culturally sophisticated, sought to renew West Lake's landscape aesthetic and engraved their poems in stone stelae at the locations of the ten classical scenes during their southern imperial tours (Wang 1999; Chen 2003; Clunas 1997).

West Lake's reputation for romantic beauty, folklore and mythology, and literati culture continues to this day and its landscape evolved into a national and international garden archetype. This archetypal spatial form was known as a water body containing the man-made elements, two causeways and three islands, surrounded by "cloud-capped" mountains on three sides (Padua 2014). Additionally, various temples, imperial gardens and natural elements on West Lake's islands, causeways, and lakefronts in the surrounding mountains are constituent components of the West Lake Cultural Landscape. West Lake's archetypal form and its ten classical scenes were replicated in other gardens and cities in China. At least thirty-six "west lakes" were known to have been replicated in Chinese cities during the Qing dynasty (Chen 2003). Scenery manipulation from the Chinese Picturesque also came into play for West Lake's *shan-shui* landscape tradition. For example, a popular literati scene or "framed" view of the lake's waterscape was created by two structures (Baochu Pagoda, *Baochuta*, and Leifeng Pagoda, *Leifengta*) on two opposing hills with the cloud-capped mountains as a backdrop.

It is important to note that while West Lake and its various imperial gardens and sacred places dispersed around the lake's perimeter were idolized and romanticized during the Classical and Later Imperial periods, it was largely the domain for emperors, the imperial court elite and religious sects. The lake was utilitarian for local fishermen and others who maintained the lake as the city's water source; but its use for leisure purposes was largely restricted to the imperial court. As mentioned in an earlier chapter, the notion of the public and leisure sphere was limited in Chinese feudal society. Typically, the local community socialized or gathered informally in temple courtyards and streets in the market district within the walled city. Additionally, periodic temple fairs during religious holidays and seasonal festivals provided places for families and friends to gather, socialize and celebrate public settings.

Many of West Lake's classical features were largely unchanged until the 20th century. However, much of the interior of Hangzhou's walled city was burned and destroyed during the Taiping Rebellion circa 1860s. Significant population decline at the time was due to high mortality rates from the Taiping Rebellion and some Hangzhou residents fled to Shanghai, one city the Taipings were not able to overcome. Additionally, as Shanghai experienced colonial modernity after the Opium Wars and its international cosmopolitan reputation grew, Hangzhou's urban reputation suffered. Hangzhou's population dramatically declined from 1 million in 1860 to 200,000 by the early 20th century (Wang 1999).

Some attempts were made to modernize Hangzhou after the first Sino-Japanese conflict (1894–1895) with the unequal 1895 Treaty of Shimonoseki opening the city to Japanese trade. However, it wasn't until after the 1911 Revolution that Hangzhou's municipal government embarked on a program of modernization. Ruan Xingji, an engineer educated during Japan's Meiji period when modernization was modeled on the west, led the transformation of Hangzhou's urban form (Wang 1999). The lakefront segment of Hangzhou's ancient city wall and adjacent city fabric were demolished and replaced with a modern city form. This included a chain of lakeside parks, *Hubin Gongyuan*, public piers, *Matou Gong*, Lakeside Street, *Hubinlu*, rational street grid and new business district for commercial and government uses (Dong 1999; Wang 1999). As mentioned in an earlier chapter, public parks and greening were important symbols of modernity during the Republic of China (ROC) era.

Shanghai's parallel colonial activities during the ROC caused explosive growth and Hangzhou's municipal government saw an opportunity to attract Shanghai's nouveau riche. City officials targeted West Lake for tourist consumption with "all kinds of new modern comforts to sell its supposed antiquity" (Wang 1999, p. 120). The Shanghai-Hangzhou railway built circa 1909 enabled the ready access to West Lake by Shanghai residents. Trees were replanted at the Bai and Su causeways and the locations of West Lake's ten poetically named scenes, areas and other locations that were cleared or destroyed during the earlier Taiping Rebellion or Japanese occupation.

To celebrate the city's new "modern" image that physically linked West Lake with Hangzhou's urban edge, modern amenities like street lighting and public promenades were built and incorporated into the lakefront. It was the first time in the city's history that West Lake was visually and physically connected to the city, essentially for consumption by the general public. New lakefront hotels were also built and subsequently private lakeside residences were built for Hangzhou's high society. New hotels were also built near the Bai causeway and some of the temples were restored (Wang 1999).

As part of the park re-naming trend and nation-building optimism following Sun Yat-sen's death, the city renamed the imperial garden on Solitary Hill, *Gushan*, near the Bai Causeway, Zhongshan Park. Hangzhou was also the host for the 1929 West Lake World's Fair. This international expo was situated along the lake's northern edge where over 20 million people visited in a three-month period (Fernsebner 2002). However, like the rest of China, the subsequent period of warlords, second Japanese occupation and World War II, and internal civil strife, ended Hangzhou's Republican-era modernization efforts.

The Mao era brought another set of changes to West Lake. Basic maintenance was needed after the civil war and communist revolution, and the city's initial activities involved dredging of the lake's bottom. The poorly maintained hillsides around West Lake were restricted to the public so that a ten-year afforestation effort with the planting of over 30 million trees could take place (Shi 2009). Like other urban locations in China, the city initiated efforts to transform Hangzhou's lakefront parks and urban form to follow the Soviet model. Green beltways and major tree planting along major roads were established in the West Lake area. Mao era changes to the Southern Scenic Area included a Youth Park (hard courts and playfields), Elder Park (senior citizens' use) and general renovation of the Orioles Singing in the Willows and adjacent imperial garden (Z. Bao 2009, pers. comm., 29). The Hangzhou Landscape Architecture Design Institute (the state-owned enterprise that later worked on the SSAMP) was established in 1952 and affiliated with Hangzhou's municipal government.

At a city public works meeting in 1953, a visiting Soviet urban planning expert was impressed by West Lake's landscape and cultural resources, and recommended focusing Hangzhou's city image on recreation, tourism and cultural activities, as well as serve as an international conference center – "Geneva of the East" (Gao 2006). In that context, the lakefront and adjacent urban zone were designated for recreational, cultural, conference hotels and public uses with industrial zones and factories located in Hangzhou's urban periphery. West Lake also served as a 1950s CPC leadership retreat and place for summer holidays; and as noted in an earlier chapter, Mao resided at one of the lakeside residential villas during his frequent visits and held meetings at the Dahua Hotel, then a state-owned property in the Southern Scenic Area (Barmé 2012). Like other places throughout China, Mao's failed economic policies, famine and poverty, and the Cultural Revolution, halted any growth in and around Hangzhou, and the area remained moribund until the late 1970s. Around this time, government officials moved into the ROC era lakefront villas and made them their permanent residences or summer homes.

Project inception

Like the rest of China, Hangzhou and the surrounding villages actively participated in the TVE program, the first wave of Deng's economic reforms, and the local economy thrived. Furthermore, people returned from the countryside to Hangzhou to find jobs in and around Hangzhou. With unregulated growth, West Lake's water became polluted from waste and sewage from sub-standard rural housing, farmers and small industries along the lake's perimeter, especially along the western and southern edges (X. Wang 2008, pers. comm., 30 June).

Subsequent to China's 1982 enactment of the Law on the Protection of Cultural Relics, West Lake was listed on China's inaugural list of forty-four National Scenic Areas, and Hangzhou, along with Suzhou and Chengdu, were on the inaugural list of twenty-four National Historical and Cultural Cities. These actions triggered a city planning process and in their first post-Mao long-range city master plan (1981–2000), Hangzhou municipal government focused on two aspects: spurring economic growth in industrial zones and protecting West Lake as a cultural resource. The latter informed their ambition to be designated a "national tourist city". In this context, economic development zones were located in the outskirts of the former walled city with the lakefront designated for recreational and cultural uses, and related commercial uses.

Soon after Deng's 1992 Southern Tour, China's State Council approved the establishment of four national-level development districts in suburban Hangzhou which allowed foreign direct investment zones. Concern about the condition of the West Lake area grew steadily in the 1990s as Hangzhou became more affluent in the wave of Deng's reforms. In the updated 1996–2010 city master plan, they expanded their identity as "an international tourist city and national Historic and Cultural Famous City, a central city in the Yangtze River Delta Region, and the political, economic, scientific-educational, and cultural center of Zhejiang Province" (Qian 2012, p. 47). Satellite zones were designated in suburban Hangzhou.

City efforts to remake themselves were ongoing and critical upgrades to West Lake remained an ongoing priority. Republican era lakefront villas were acquired from CPC officials, and vacated and renovated by local government in the 1990s. Some were re-purposed for commercial, cultural and leisure uses. The most significant move was the demolition of existing walls enclosing the private gardens of these lakefront villas. This expanded lakefront use increased vegetated public open space and the city established "free" access to a continuous lakefront public landscape that required no entry fees or admission charges. Demolishing garden walls to increase public accessibility and vegetated lakefront public spaces represented the city's priorities to renew and reclaim their "*shan-shui*" West Lake cultural identity and modernize Hangzhou.

However, the environmental problems associated with the lake continued and were exacerbated by the area's urbanization and increasing tourist population. Additionally, West Lake's public infrastructure had not been fully modernized and hindered foreign direct investment in the suburban satellite special economic zones. New and enlightened city and provincial leadership emerged in late 1998. Their sophisticated agenda specifically embraced urbanization, promoted globalization and envisioned West Lake as a cultural resource for the global common good.

Hangzhou's new leadership prioritized the agglomeration of compatible land uses including high technology zones linked with university expansion, in addition to the notion of West Lake as a cultural resource for the global common good. In effect, historic Hangzhou was formally incorporated into West Lake's cultural protection zone and contributed to local, provincial and national agendas focused on achieving the prestigious UNESCO designation as a World Heritage site. New government agencies were created to assist with the new leaders' city-making ambitions including the Hangzhou Economic and Technological Development Area Administration Commission (oversight and supervision of activities in Hangzhou's development zones) and Hangzhou West Lake Scenic Area Administration Commission. The latter agency is a cooperative effort between the city's Bureau of Gardening and Cultural Relics and Hangzhou West Lake Scenic Area Committee for activities related to the protection and management of West Lake as a national-level and international cultural resource. Hangzhou's new leaders, like their earlier Republican era counterparts, sought to capitalize on China's growing *nouveau riche* and consumer class, especially in nearby Shanghai. By the turn of the 21st century, leisure time had increased and restrictions for intra-provincial travel were relaxed. Furthermore, national capital projects to link Hangzhou to other major cities by road and rail were also underway at the time (Ma & Wu 2005).

A key government initiative circa 1999 called for an in-depth environmental assessment of West Lake with the final report known as the West Lake Comprehensive Protection Project (Project), *Xihu Zonhe Baohu Gongcheng*, and set the framework for HDI's work on the SSAMP project. The primary goal that emerged out of this larger comprehensive project was to reclaim West Lake's iconic cultural landscape identity while regenerating Hangzhou for the new century. In this light, the city sought to reimagine the city's West Lake's late imperial image: "one lake, two causeways, three islands, two pagodas, *yi hu, er ti, san dao, er ta*", with a contemporary image of the lake defined by "hot east, prosperous south, secluded west, elegant north, and beautiful center, *dong re, nan wang, xi you, bei ya, zhong liang*" (W. Zhou 2007, pers. comm., 7 May). Eventually, the city's Bureau of Gardening and Cultural Relics launched a design competition that the Atelier DYJ/HDI consortium won and the outcome of this project outlined a broad comprehensive renovation strategy for West Lake with HDI's scope of work focused on "Detailed Planning and Design". Parallel efforts were also underway and involved eliminating pollution sources, sub-standard housing, pig farms and various nonconforming land uses along West Lake's western perimeter, and relocating the affected population.

Project realization

The city approved various West Lake redevelopment projects in early 2002 including the necessary public infrastructure projects critical to rejuvenate West Lake's eastern and southeastern edge or Southern Scenic Area. The "West Lake South Periphery Integration Project, *Xihu nanxian zhenghe gongcheng*", was one of these projects. This project included the 2002 construction of key design components from the SSAMP by HDI. HDI's work focused on the construction of a new lakefront pedestrian path, *binhu jingguan dai*, along the eastern and southeastern shores of the lake that also linked the system of public parks internally and externally with the urban activities along Hubin and Nanshan Roads. HDI's design aligned with the city's objective to eliminate entry fees and ticket sales for a new official "Lakeside Scenic Belt, *Hubin Jingguan Dai*", a scenic corridor that integrated Southern Scenic Area's lakefront path and series of public open spaces with the tree-lined cultural and commercial districts along Nanshan and Hubin Roads (Shi 2009). HDI's work also aligned with the city's goal to reimagine the popular expression for West Lake, "misty, cloud-capped hills on three sides with a city on the fourth, *san mian yunshan yi mian chen*".

Deputy Mayor Zhang saw the potential for tourism from nearby Shanghai, post-socialist China's designated international financial center, and visitors traveling by private car and train from other parts of the nation. Given the new leaders' vision of West Lake as a cultural landscape commodity for the global common good, and ambitions to gain UNESCO world heritage status, HDI drew their design inspiration from West Lake's cultural heritage (folklore, mythology, imperial history, literati arts and Chinese Picturesque) and Hangzhou's urban heritage (ancient, imperial and 20th century). HDI also acquired oral histories from local senior citizens to get a sense of the lakefront uses before the Japanese invasion and World War II (W. Zhou 2007, pers. comm., 7 May).

While the city was attuned to West Lake as a major tourist designation and cultural resource, HDI's design objectives focused on the cultural landscape's visual quality and pedestrian amenities, especially visual and physical linkages to West Lake from the adjacent urban and cultural districts. More importantly, HDI sought to create a didactic experience where visitors learned about local urban history, the ancient and imperial history of Hangzhou, as well as West Lake's folklore and mythology.

HDI's detailed design for the SSAMP, an area over 300 hectares (740+ acres) was informed by various activities: archival research; environmental analysis, especially water quality; in-depth studies of cultural resources, as well as detailed site analysis. HDI reviewed historical records on dredging and management activities for the lake and the city's historical development including cultural and religious phenomena. HDI's review of visual archives gave them a deeper understanding of ways that West Lake and its urban edge were portrayed in paintings during various dynasties. HDI also utilized the internet as a resource for design inspiration and reviewed websites of the leading international landscape architects of that time.

HDI's research revealed that the walled city's lakefront had contained natural creeks and inlet locations for canals that supplied the ancient and imperial city with water. They also found evidence of imperial gardens and boat docks outside the city walls, as well as the different paths to and from the city's gates used by the emperor and his imperial court or others. They confirmed Hangzhou's modern gesture from the Republican era and the creation of the first unobstructed water-mountain view and public open space experience through the demolition of the city's ancient walls fronting West Lake. They also confirmed the early 20th century establishment of a series of lakeside public parks with a pedestrian promenade and public boat piers. West Lake's original Ming and Qing dynasty waterfront alignments were altered due to silting and modernization projects in the earlier 20th century Republican era and mid-century Mao period. HDI's design intention was to reclaim the Ming-Qing lakefront alignment and urban canal locations.

In broad strokes, HDI's detailed planning and design focused on four dimensions: 1) West Lake's landscape legacy; 2) cultural and urban heritage; 3) human-scale, culture-based lakefront experience; and 4) natural resources, especially the water system (lake water quality and stormwater management). Articulation of the SSA lakefront was reimagined to consider archival records of dredging, the city's canals, and representations of the lakefront in poetry and paintings. The dynamic inter-relationships of the lake's water quality, surface water drainage and best stormwater management practices, and long-term maintenance of the lake helped to inform HDI's decisions on SSA's spatial organization and final waterfront alignment. HDI's design approach erased the Soviet design influence and remade the image of these former Mao era sites to reflect their understanding of contemporary needs. Integrated with these culture-based and ecological design strategies, HDI's contemporary design (Figure 5.44) created a 21st century contemporary park experience for consumption by local, national and international tourists.

HDI's completed SSA projects included First Lakeside Park, *Hubin Yi Gongyuan*, Yongjin Park, *Yongjin Gongyuan*, and Yongjin Square, *Yongjin Guangchang*, Orioles Singing in the Willows, *Liu Lang Wen Ying*, and Long Bridge Park, *Changqiao Gongyuan*. Other completed works included the new construction of Jixian Pavilion, originally built in the Qing dynasty to attract scholars back to West Lake, and reconstruction of the King Qian Memorial Temple complex. HDI replaced the Mao era parks, effectively erasing the memory of Soviet-influenced park design. The Mao era "Elder" (senior citizens) park was replaced with the commercial project, *Xi Hu Tiandi*, West Lake Paradise, a new restaurant/retail complex modeled on Shanghai's *Xintiandi* (literal translation is new heaven on earth or new paradise), a pedestrian-oriented commercial building complex that combined the adaptive re-use of historic buildings with new buildings. Benjamin Wood, an American architect based in Shanghai, designed

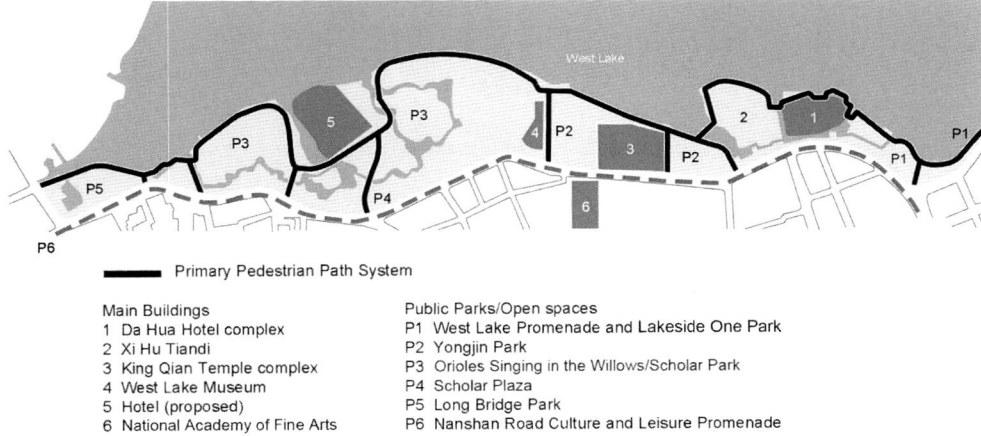

Figure 5.44 SSA Plan of design components based on HDI original, published with permission, graphic by J. Wu

the Xi Hu Tiandi building complex that combined new glass buildings with the renovated Republican era historic buildings nested within gardens designed by Design Land Collaborative, a Shanghai-based firm founded by American landscape architect Dwight Law (B. Wood 2019, 15 June correspondence). HDI's work focused on the lakefront promenade at Xi Hu Tiandi. Adjacent to Xi Hu Tiandi is the Dahua Hotel (circa 1930s) where HDI renovated the gardens with a new lakefront promenade and new contemporary "zig-zag" stone bridge that traversed the hotel's West Lake frontage.

HDI's SSAMP integrated streetscape improvements along Nanshan Road, a designated "Culture and Leisure Promenade Corridor", especially given the new primary building façade for the National Academy of Fine Arts was just across the street from the Southern Scenic Area. In addition to streetscape amenities, HDI's completed projects along Nanshan Road included Yongjin Square, a contemporary plaza at the site of the historic Yongjin gate of the ancient walled city and access point to Xi Hu Tiandi; a new gateway to the Oriole's Singing in the Willows park; a contemporary Song-style colonnade that replaced a Republican era neo-classical colonnade, the so-called Roman gate to a Soviet-style Youth Park (recreational courts and playfields) that was left to deteriorate in the Mao era; and an entry plaza to Long Bridge Park.

SSA experience

The following experiential narrative summarizes HDI's completed work on nine lakefront components and three key projects along Nanshan Road and discusses as three spatial clusters: 1) a series of six connected public open spaces; 2) three lakefront public open spaces; and 3) three separate elements along Nanshan Road. Like the previous three case studies, this narrative is based on the designer's intentions garnered through archival review of original design documents, interviews and personal correspondence, field research and highlights of the experiential qualities. The Southern Scenic Area represents HDI's contemporary interpretation of West Lake's landscape legacy combined with Hangzhou's urban heritage through a complex design influenced by both international and Chinese Picturesque design language. This fourth case study narrative illustrates the key dimensions that characterize hybrid modernity: fusion of local and international design influences, local identity and nation-building.

Lakeside One Park, Da Hua Hotel, Xi Hu Tiandi, Yongjin Park, Yongjin Square and King Qian Memorial Complex

Lakeside One Park is one of five lakefront public parks created in the early 20th century Republican era when the city's modernization established for the first time a physical and visual link between West Lake and the city. HDI's work included a new continuous lakefront path and extension to the adjacent Lakeside Two Park and public lakefront promenade path along the SSA, renovation and adaptive re-use of Republican era buildings, entry plaza along Hubin Road, reconstruction of Jixian Pavilion, and additional new design elements and amenities. The width of the new contemporary promenade along Lakeside Road, *Hubinlu*, accommodates pedestrians, bicycles and small motorized shuttles.

Cut stone in a rectilinear paving pattern defines the waterfront promenade path throughout the length of the SSA. Along linear waterfront segments of this path at the Lakeside One Park extension, HDI incorporated a wide-stepped edge where large groups of visitors could gather and enjoy the dramatic panoramic waterscape view with the mountain backdrop. The promenade design incorporated other amenities: smaller covered sitting areas defined by a contemporary pavilion design with linear benches and an overhead canopy oriented to panoramic views of West Lake. One highlight is a large field of ground occupied by a representational relief map of Qing dynasty Hangzhou. It illustrates the city's canal and street circulation system, outer walls and gates, inner walled imperial palace, other major city elements and the landscape outside the walls including the West Lake frontage; and its design allows visitors to walk on it and around it. During weekends, the author observed families walking on the historic map, pointing to city elements and discussing the walled city's urban form. Near the map, HDI's promenade design incorporated another didactic design element: a rectilinear interactive water feature. This water feature, made of stone and metal, is designed with contemporary elements that symbolize Hangzhou's historic canals and water wells. It is not restricted by fencing and visitors can touch water flowing out of the symbolic canals. Bilingual (Chinese/English) interpretative signage on a raised stone plinth explains the fountain's symbolism and that lake water was the historical potable source for consumption. Walking southward, the lakefront promenade makes a slight curve that parallels the bend in the adjacent Hubin Street where it changes to Nanshan Street. The promenade opens onto an entry lakefront plaza that is also the terminus of Youdian Street. The plaza defines a gateway and transition from the city to the Southern Scenic Area. Within thirty paces, the promenade view shifts to a layered scene: a foreground view of a field of floating lotus plants in West Lake, middle ground view of the Qing style Jixian Pavilion and panoramic water-mountainscape background view (Figure 5.45). This dramatic West Lake view is accentuated before and after sunset, and the experience sets the stage for the visitors' initial impression of West Lake.

Jixian Pavilion is a reconstruction, originally built in the Qing dynasty to lure artists and poets back to Hangzhou. It is a major cultural icon and visual focal point on a hexagon-shaped stone platform at the end of a linear pier (Figure 5.46). The pavilion

Figure 5.45 Layered view with Jixian Pavilion in the middle ground and panoramic background view of West Lake and the mountains, photograph by author

Figure 5.46 Reconstructed Jixian Pavilion marks the northern end of the Southern Scenic Area, photograph by author

allows up to twenty visitors to experience West Lake's panoramic waterscape view. From the pavilion's vantage point, visitors can experience framed views with urban and natural edges, as well as views of the tree-lined Su Causeway, Solitary Hill and distant mountainous backdrop.

Southward along the promenade are several mature trees, remnants of Republican era lakefront residences. These trees were nested in a new lakefront sitting area among renovated historic buildings converted for commercial use. Moving southward, other Republican era mature trees mark the southern edge of Lakeside One Park and transition to the Dahua Hotel complex. The vista opens along the promenade as visitors cross an arched stone bridge suspended over an artificially created canal that spatially defines the circa 1930s Dahua Hotel and building complex.

HDI's work on the Dahua Hotel complex consisted of upgrading the grounds and creating a new publicly accessible promenade along the property's lakefront that connected to Lakeside One Park at the north end and Xi Hu Tiandi, the commercial complex, on the south end. HDI's major design move was the insertion of a new contemporary "zig-zag" stone bridge (Figure 5.47) separated from the hotel property and set in the water along Dahua Hotel's lakefront. Zig-zag bridges are Chinese Picturesque elements typically utilized for scenery manipulation. In this context, visitors can experience changing vistas of West Lake's water-mountain scenery while slowly walking along the stone zig-zag bridge. At the bridge's southern point, it makes landfall at a lakefront terrace with food concessions and areas for sitting. The lakefront promenade path continues for a short distance when it transforms into another arched stone bridge that crosses a canal lined with willow trees. It mirrors the historic city canal alignment and defines a lakefront entry to Xi Hu Tiandi, a commercial restaurant development on the former Mao era park for the elderly, mentioned earlier.

HDI's work on the waterfront edges of Xi Hu Tiandi's fifty-hectare (123-acre) commercial area demonstrates a high level of craft and attention to design detail. The lakefront path was similar in style to Lakeside One Park and the Dahua Hotel. Their work on the promenade path wrapped around the re-purposed Republican era buildings fronting West Lake (Figure 5.48) and contiguous waterfronts of Yongjin Pond and Yongjin Square at the commercial's southern edge. Wood's new glass retail and restaurant buildings in Xi Hu Tiandi were situated along the frontage of Yongjin Pond. Apparently, glass buildings at the time were considered temporary and Wood was on an expedited schedule to meet the city's deadline (Wood 2019 correspondence). Unlike Shanghai's Xin Tiandi dense urban setting, Hangzhou's Xi Hu Tiandi's concentrated complex of buildings within a garden setting is nested within a public park (Figure 5.49).

Visitors strolling the lakefront promenade from Xi Hu Tiandi have the choice of continuing southward and crossing the Yongjin Pond along a curving sixty meter (180 feet) bridge to Yongjin Park. Alternatively, visitors can walk along the pond's waterfront to Yongjin Square and into the city. Yongjin Square marks the location of the walled city's Yongjin gate, *Yongjinmen*. Yongjin Pond (Figure 5.50) is a reimagined natural stream that traversed into the historic city where it transformed into the city's internal canal and transportation system. Yongjin Square also marks a gateway from the city to the SSA and Xi Hu Tiandi. This contemporary plaza consists of a field of cut stone paving with a stepped waterfront edge. It overlooks Yongjin Pond, which contains two figurative sculptures that reflect local heritage: a golden water buffalo

Figure 5.47 Zig-zag bridge along Dahua Hotel's frontage extends the West Lake promenade experience and creates various views of West Lake and surrounding scenery, photograph by
author

Figure 5.48 HDI's waterfront promenade and renovated historic building by Benjamin Wood in
Xi Hu Tiandi commercial area, photograph by author

Figure 5.49 Glass buildings in Xi Hu Tiandi by Benjamin Wood and gardens designed by Design Land Collaborative, photograph by author

Figure 5.50 Reconstructed Yongjin pond with two sculptures representing local folklore and mythology, photograph by author

Figure 5.51 Golden water buffalo, mythical water source for West Lake and pedestrian bridge that links Xi Hu Tiandi and Yongjin Park, photograph by author

(Figure 5.51) wading in the water represents West Lake water source mythology; and 2) Zhang Shun, a hero from a Ming dynasty novel who was killed in the water outside Yongjin gate.

Walking southward from the urban-oriented Yongjin Square, the scenery shifts to a more park-like experience as visitors enter Yongjin Park. The lakefront promenade path transforms into a greenway with willow tree plantings set in lawn and along the lakefront edge. Amenities include contemporary wooden benches and lighting. The path widens where visitors can decide to visit the King Qian Memorial complex, a series of replica ancestral buildings and museum set back from the lakefront (Figure 5.52), or continue into the park. The new replica complex houses cultural artifacts and celebrates the area's contributions when it flourished as a capital city for the Wuyue kingdom. The new memorial complex was the third reconstruction on a former Qing dynasty location. It was ravaged during Mao's Cultural Revolution and considered uninhabitable. The complex is set back from Nanshan Road and spans the area to West Lake where a symbolic waterfront stone platform marks the historic location of a pier used by emperors and kings.

Orioles Singing in the Willows/Scholar Park, Long Bridge Park

The site of the royal pier marks the change and transition to the Orioles Singing in the Willows/Scholar Park. Named for one of West Lake's classical scenes, the lakefront

Figure 5.52 Gateway to King Qian Memorial building complex along the lakefront, photograph by author

promenade transforms with a lush planting of willow trees. A secondary path leads inland to the West Lake Museum and the primary willow-lined lakefront path traverses southward deeper into Orioles Singing in the Willows, also nested within Scholar Park. This area has been documented as the largest imperial garden outside the walled city and within proximity of the gate near the hilltop imperial palace; and it was a place where the emperor would stroll after being carried from his palace in a sedan chair (W. Zhou 2007, pers. comm., 7 May). Along the willow-lined lakefront path are reconstructed Qing style pavilions where visitors can sit and enjoy panoramic lake-mountain scenes framed by the reconstructed Leifung Pagoda on the southern hilltop. Secondary paths within this passive area "twist and turn" through a Chinese Picturesque landscape with reconstructions of historic pavilions, stone and paths bridges that meander around artificially constructed streams and ponds.

The willow tree plantings along the lakefront path become less dense as visitors walk southward. The widening of the lakefront path with much fewer trees marks a transition to Long Bridge Park. This park's experience is centered on a new 150 meter (450 feet) long contemporary style bridge (Figure 5.53). It celebrates a romantic folk tale about a young couple who commit suicide. The contemporary stone bridge set back from the land's edge contains long segments barely raised above the lake water level with a few "zig-zags". These long segments and path orientation create a floating experience with changing views of the surrounding mountain-water scenery. The bridge widens to create a large platform at one of the angle changes where large groups can gather to contemplate the lake's waterscape view framed on the southern edge by the Leifung Pagoda; a few zig-zags beyond is a covered traditional waterfront pavilion with framed views of West Lake and another view of the pagoda (Figure 5.54). Where

Figure 5.53 Long Bridge defines another park at the southern end of the Southern Scenic Area, photograph by author

Figure 5.54 Traditional pavilion connected to the contemporary stone Long Bridge with framed views of Leifung Pagoda and West Lake, photograph by author

Figure 5.55 Long Bridge park entry at tree-lined Nanshan Road, photograph by author

the bridge lands at the lakefront promenade, visitors can continue walking southward along the tree-lined lakefront path in Long Bridge Park, a contemporary Chinese Picturesque park, or walk a few paces east to Nanshan Road, a tree-lined cultural district (Figure 5.55).

Nanshan Road streetscape and three major design elements

Aside from Yongjin Square, an urban gateway to Xi Hu Tiandi, HDI's work along Nanshan Road incorporated streetscape upgrades (vegetation and paving) and three major design elements: Scholar Plaza; the gateway to Orioles Singing in the Willows; and the King Qian Memorial Archway. The Scholar Plaza replaced a former site of a Republican era neo-classical U-shaped colonnade that marked the entry, a so-called Roman gate, to a Mao era youth park (hardcourts and playfields). The colonnade was considered too expensive to demolish in the 1950s and 1960s Mao era, and left to deteriorate. HDI replaced the degraded colonnade with a new structure (Figure 5.56) nested in a new public plaza along Nanshan Road. The new colonnade followed the Republican era U-shaped alignment. However, the replacement column and base structure represented a traditional Song era building and its overhead lattice represented HDI's contemporary interpretation of Song era architecture (W. Zhou 2007, pers. comm., 7 May). The re-imagined plaza incorporates a curvilinear shaped reflecting pool that recalls the memory of a natural stream that historically occupied this area. Like HDI's other new waterfronts, the plaza's design integrated steps at the pool's edge. The plaza's contemporary paving design incorporated a rectilinear patterned

Figure 5.56 HDI's Scholar Plaza design incorporates a contemporary reflecting pool and paving with a new colonnade made of Song-style columns, replacing a neo-classical Republican era structure, photograph by author

grid of cut stone and traditional building materials. HDI intended to preserve the site's memory of the Republican and Mao eras by retaining the colonnade's original curvilinear form (W. Zhou 2007, pers. comm., 7 May). Many families frequented this square with children playing in the pool during weekends.

HDI's design of the Nanshan Road gateway to the Orioles Singing in the Willows, one of West Lake's ten scenes that refers to the coming of spring, incorporates three sequential design elements: nine-meter (thirty-feet) wide square of cut stone paving with an intricately carved circular phoenix (square symbolizes heaven, the phoenix and circle are symbols of the emperor) medallion at its center; pavilion gateway structure; and intricately designed tree-lined path. The special "circle in square" paving design (Figure 5.57) is referential to the emperor and imperial times and marks the major gateway to the SSA at Nanshan Road. It also marks the initial entry sequence to the Orioles Singing in the Willows and traditional garden. Spanning the entry is a large Chinese-style non-descript pavilion with the four-character scene inscription overhead, the second component of the visitor's three-part entry experience. Walking through and under the pavilion creates a transition from Nanshan's urban experience into a passive park. The pavilion emphasizes the change of scenery, and the framed axial path view accentuates the path's fifty-meter (165-feet) length. HDI's elegantly executed axial path is a contemporary design that reimagines an imperial walkway.

The symbolic imperial path contains a central stone path flanked by two double rows of willow trees (Figure 5.58) and leads to one of West Lake's ten poetically

Figure 5.57 Special paving and gateway marks the entry to SSA and the Orioles Singing in the
Willows, one of West Lake's Ten Scenes, photograph by author

named scenes, Orioles Singing in the Willows. Along the center of the central stone
path is an articulated ornamental stone band that follows the aesthetics of the imperial phoenix paving at Nanshan Road. Willow trees flanking the central stone path are
set in a field of beige-colored cobbles with two wider paths on either side for larger
groups of visitors. Large boulders are artfully set into these wider paths and provide
places to sit, and are used as an edge for a corridor of ornamental plantings. On the
one hand, this entry sequence symbolizes a path for emperors and his court. From the
perspective of a western-educated designer, critic and contemporary historian, HDI's
contemporary axial path design creates a serene and elegant experience. Its characteristics are similar to the tree allée or *bosquet* perfected by Andre Le Notre's French
royal gardens or formality of tree plantings by Dan Kiley in his "classical modern"
works. The craftsmanship of HDI's axial path demonstrates exquisite craftsmanship
and attention to detail.

The third design element of note along Nanshan Road is another axial design by
HDI. It consists of a series of five equally spaced archways, *pailous*, with traditional
roofs (Figure 5.59). These define a gateway and pedestrian path to the King Qian
Memorial Complex and West Lake's waterfront. Each archway commemorates one
of the five kings who ruled during the Five Dynasties period. It also represents the
symbolic sacred path utilized at burial sites for Chinese kings and emperors. The path
experience is monumental in comparison to the Orioles Singing in the Willows axial

Figure 5.58 HDI's axial stone pathway and double allée of Willow trees symbolizes a royal path and demonstrates exquisite design and attention to detail, photograph by author

Figure 5.59 Five equally spaced archways commemorate Hangzhou's five ancient kings and mark the path from Nanshan Road to West Lake's waterfront and King Qian Memorial complex, photograph by author

path. The archways are nine meters (thirty feet) wide and nine meters high and lined by formal tree and shrub plantings in the lawn with a cypress hedgerow taller than the archways. The rhythm of walking under and through the archways are accentuated by the soft green wall of the tall cypress hedges. In addition to creating an experience that anticipates the iconic view of West Lake, this formal axial path along a green corridor allows visitors to reimagine ways that the emperor and his court would pay homage to the region's ancestors and royal heritage.

Hybrid modern synthesis

The most striking features of HDI's work are ways that international design influences and the Chinese Picturesque genre are interwoven to create a culture-based leisure experience without the feeling of a Disney theme park. The various designed landscapes within the Southern Scenic Area are outdoor settings where visitors can re-imagine West Lake's landscape aesthetic in contemporary or historic terms. The major bridge elements (zig-zag bridge at the Dahua Hotel complex, expanded linear bridge traversing Yongjin Pond and Long Bridge at the southern terminus) emphasize West Lake's mountain-water landscape and the visitors' contemporary leisure experience. The utilitarian and imperial connections between Hangzhou and West Lake are artfully expressed in HDI's design narrative through water features, sculpture and re-articulated lake edge. The adaptive re-use of structures from the Republican era, the use of figurative sculpture and other deliberate design elements are referential to local folklore and mythology.

The most dominant element, next to the focus on West Lake, is the way that HDI deploys water. Symbolic shapes and forms were referential to the city's natural and urban history (streams, bays or ponds and canals). HDI re-shaped and re-articulated the lakefront's alignment to represent their interpretation of Qing dynasty paintings of West Lake, and HDI were also influenced by Ming Dynasty poetry about West Lake. Not only was water utilized for symbolic design gestures, HDI's use of canals contributed to their ecological design approach and scientific understanding of maintaining the lake's water quality. Table 5.4 summarizes HDI's various design approaches and hybrid modern design language.

Three fundamental elements of the hybrid modern design grammar are used unselfconsciously in the SSA: references to site history and local history, concepts and techniques influenced by recent developments in international design, and the design grammar of the Chinese Picturesque. The design approach was strongly influenced by site history and went to some length to incorporate elements of local history. This was done through means that ranged from the re-use of existing structures to symbolic references to ancient features of the area.

HDI's design also shows the influence of recent trends in park design outside China. Cultural heritage preservation, ecological design, didactic or interpretative, *genius loci* and spirit of place, site history and symbolic references to the local cultural context have been among the distinguishing features of contemporary landscape design. Finally, the design also employed virtually every convention used in the design language of Chinese Picturesque – including scenery manipulation, i.e. framing and borrowing scenes along with "twists and turns" in secondary paths, and shifting waterscape scenes while walking along zig-zag bridges; the use of rockery; waterfront pavilions and the like. HDI also constructed canals to re-articulate the

Table 5.4 Summary of case study four: SSAMP's Hybrid Modern design language, by author

Design approach	Design grammar	Design Vocabulary	Materiality
Genius Loci (Spirit of Place) Cultural	Reclaim the site's cultural history	Folklore & mythology	Lakefront path, figurative sculpture; waterfront pavilions; Long Bridge
		Ancient kingdom, imperial, literati and 20th century	Reconstruction of King Qian Memorial complex, imperial boat pier; re-articulation of lakefront alignment mimics Qing/Ming dynasties paintings and poetry; Orioles Singing in the Willows, Scholar Garden, Scholar Plaza, Jixian Pavilion, Willow tree plantings
Urban	Symbolic	Historic walls + gates	Historic renovation and re-use of Republican era buildings and public boat piers
		Water supply	Lakeside One Promenade: Qing city map paving design, fountain design represents Hangzhou's historic canals & wells
			Yongjin Pond former site of natural stream that functioned as a city water source; Yongjin Square marks historic location of walled city gate
			Axial path to Orioles Singing in the Willows reimagines royal path to poetic scene
		Republican era	Scholar Plaza: colonnade retains memory of Republican era with U-shape form, reimagines new colonnade with Song dynasty columns and entry to imperial garden; reflecting pool symbolizes former stream outside historic gate; contemporary paving utilizes materials used in local traditional building
			Remake Lakeside One Park preserves memory of Republican modern city-making and inaugural West Lake physical and visual city links with a series of lakefront public parks along the alignment of the demolished historic city walls
			Adaptive re-use of lakefront villas preserves memory of the 20th century Republican era
Didactic	Teach local cultural and natural history	Experiential nodes; Historic place-markers; Visual focal points; and lakefront path	Qing city map, fountain design represents canals and wells; Jixian Pavilion preserves Qing dynasty urban memory to attract literati; renovated early 20th century lakefront villas + gardens preserve Republican era Orioles Singing in the Willows "poetic scene" preserves literati + imperial culture
			Nanshan Road gate with imperial willow-lined axial path creates new imperial experience
			Golden buffalo + hero sculptures in Yongjin Pond reaches local folklore and literature; Yongjin Pond marks location of natural stream and water supply for city; Yongjin Square marks location of walled city gate; Scholar Plaza marks historic location of imperial path to West Lake scenic garden from royal palace in the walled city and contains a new U-shaped colonnade reimagines local building tradition with Song style columns; and the U-shaped alignment retains the memory of the Republican era neo-classical colonnade
			King Qian Temple memorial complex + archway axial path preserve local ancient history and the memory of former kings
			Long Bridge reimagines romantic folklore
			The new gardens and waterfront parks within the Southern Scenic Area reimagines royal and literati elite culture

(Continued)

Table 5.4 Summary of case study four: SSAMP's Hybrid Modern design language, by author (*Continued*)

Design approach	Design grammar	Design Vocabulary	Materiality
Ecological	Reclaim natural lakefront edge and natural ecology Enhance lake water quality, stormwater management	Reconstruct bays, islands, natural streams New canals, re-articulated lakefront edge and landforms	Dredging and reconstruction of lakefront edge, Yongjin Pond, Xi Hu Tiandi, Long Bridge Park Dahua Hotel complex – canals create island form Xi Hu Tiandi – canals and ponds re-articulate peninsula form Wetland plantings at artificial streams and ponds, and along waterfront edges in various locations Indigenous woodland planting
Chinese Picturesque	Mountain, water *Shan-shui* landscape Scene-making Scenery manipulation	Lake and mountain views Borrowing scenes Framing scenes Scene definition	Strolling along continuous lakefront path creates a variety of scenes that are revealed while in motion Pavilions to look from, look through, and look at from a distance. Lakefront stationary locations, i.e., sitting areas with mature tree plantings and pavilions create panoramic *shan-shui* views and a variety of framed scenes and borrowed scenes Arched and zig-zag water bridges slow pedestrian mobility and enable shifting and constantly changing views of waterscape scenes and borrowed surrounding mountain scenery Twists and turns in secondary paths made with local stone and building materials in Chinese Picturesque gardens with new water bodies nested within SSA create various garden scenes
Nation-building	Remake West Lake Cultural Landscape as an iconic destination for the global common good and regenerate the earlier 20th century Republican modern link between the city and the lake	Fuse ancient, imperial and literati traditions with ecological design for improved water quality and local natural environment	In the contemporary reimagination of West Lake and Hangzhou in the New Era, all vestiges of Soviet style open spaces built during the Mao era were erased and replaced with references to local mythology, folklore, royal and imperial literati culture and Hangzhou's modern and historic urban heritage and relationships to the lake

lakefront and create an island for Xi Hu Tiandi, their own version of an "ecological island", to enhance water quality. The techniques are taken out of their traditional context and they often co-exist with contemporary design; but they nonetheless send a signal that the hybrid modern design is authentically Chinese when seen in the contemporary context.

SSA's expression of hybrid modern design evolved from locally trained designers. When HDI was commissioned to environmentally assess West Lake, construction on Living Water Park, Zhongshan Shipyard Park, and Jinji Lake was completed or near completion. HDI embraced the importance of ecological design from these parks and understood the significance of West Lake's natural and urban hydrologies for environmental health. The didactic experiences within SSA were tied to West Lake's long landscape heritage and Hangzhou's urban heritage. HDI employed the Chinese Picturesque design language in traditional and contemporary ways throughout their work. HDI and Hangzhou city officials were clear about West Lake's cultural heritage and the development of the SSA public park as a destination for local and national tourism, and the global common good. HDI also took advantage of the internet and world wide web and perused the websites of international design firms. HDI's work contributed to the area's overall cultural protection and laid the groundwork for the 2011 designation of the West Lake Cultural Landscape of Hangzhou as a UNESCO World Heritage site.

Four public parks – the schema for post-Mao China's hybrid modernity

The logic for the development of each of the four parks varied some. However, representations of local identity and nation-building in each park suggest common ground. The idea to develop Living Water Park was imbedded in a grassroots-driven effort to clean up the environmentally degraded Funan river system through a modernization project in Chengdu. Betsy Damon's provocative curated art events coincided with the park-building component of the city's modernization and river clean-up project that sought to celebrate local efforts and environmental leaders, and restore the river as Chengdu's definitive city icon. Zhonghsan's mayor demanded design innovation to create a public park in the spirit of the city's namesake (Sun Yat-sen) and national hero's hometown, and symbolized the city's progressive ideology. Jinji Lake's system of public open spaces was seen as critical infrastructure for Suzhou's new community identity and national development model. Revealing and reclaiming West Lake and Hangzhou's cultural heritage, including urban history (ancient, imperial and modern Republican era) was seen as critical by new leaders to regenerate the city's image, especially for the global common good. Within these broader civic terms, these four purpose-built parks were seen as major agents for cultural change, regenerating local identity and nation-building.

Each city had its own particular local identity and the designers for each of these destination or landmark parks would appropriate these cultural references within their park designs. Local identity in Living Water Park took several forms. The park's environmental theme and river water cleansing demonstration enhanced the city's local identity as ecologically progressive and launched Chengdu to the forefront of China's environmental movement (X. Sun 2008, pers. comm., 18 June). Local identity at Zhongshan Shipyard Park was derived from the site's socio-cultural history, as well as the city's namesake as the birthplace for Sun Yat-sen. Jinji Lake, a contemporary identity for New Suzhou, originally a joint venture between Singapore and China, was critical to its economic viability as a sustainable national model. West Lake was integral to Hangzhou's identity and the new design for the Southern Scenic Area would help rejuvenate that image. Revelations of the site's cultural heritage and urban legacy were essential for the remaking and rejuvenation of Hangzhou's contemporary identity. Although Hangzhou is the only place where Mao era parks were replaced with new spatial forms, regenerating and remaking local identity was a common characteristic revealed in the analysis of the four purpose-built parks.

Identity was also manifested through symbolism in the four purpose-built parks. Symbolism varied from park to park, and represented both local and cultural identity as it relates to a post-socialist society. The two-dimensional plan drawing for Living Water Park was construed as a fish by the client – a symbol for life and fertility in Chinese culture. Historically, the Funan River system, part of the Min River and a tributary of the Yangtze River, was considered a local deity. The Funan was also the source of the luster in the local silk brocade, a major industry when Chengdu was the ancient capital city of the Shu Kingdom. The riverfront was also a place for cultural productivity where poets found inspiration. LWP's artificial mountain symbolized nearby Mt. Emei, one of China's four sacred Buddhist mountains. LWP's riverfront incorporated vegetative stepped edges inspired by China's iconic rice terraces. The designers for LWP used symbolism in a multi-faceted way – local culture and identity, historical references, and China's agrarian heritage.

Symbolism at Zhongshan Shipyard Park paid homage to the Mao era – celebrating the work unit and site's industrial heritage, and creating a public experience in the Red Box to contemplate the Cultural Revolution. Jinji Lake's new waterscape symbolized historic Suzhou's eastern expansion area and new residential community and high technology zone for local and foreign professionals. The Southern Scenic Area symbolism represented West Lake's cultural prominence in local history from ancient times to the Republican period, as well as its links to Hangzhou's urban history and new post-socialist image.

Nation-building, while distinctive for each of these parks, was prominent. LWP is considered a hallmark for ecological design and environmental education, and the first of its kind in China (X. Sun 2008, pers. comm., 18 June; Z. Bao 2009, pers. comm., 30 April). ZSP, a post-industrial park, is also the first of its kind in China. It celebrates both the Mao era and China's post-socialist society with its transformation from industrial use to a contemporary new park that preserves the site's memory as a shipyard factory. It advocates cultural awakening and teaches visitors about the site's ecology and natural resources, especially the tidal dynamic relationship between the lake and adjacent river and the importance of preserving mature trees through the creation of an artificial island. ZSP also introduced native wetland plantings in a place known for ornamental horticulture. Vegetative hedgerows were spatially organized to mimic the shipyard factory's residential dormitory rooms. Jinji Lake's system of new public open spaces were core infrastructure components for the Sino-Singapore cooperation zone in SIP that set the foundation for this new town model prototype. HDI's work on the SSA enhances West Lake's national park stature, celebrates Hangzhou's urban heritage, and carries out the city leaders' vision to remake its identity for the global common good. This points to a shared pattern of individual and collective nation-building among the four case studies.

All four case studies incorporated the Chinese Picturesque design language. Some were incorporated as afterthoughts as in the case of LWP. For Jinji Lake and the Southern Scenic Area, concepts of the Chinese Picturesque were critical influences.

In the case of ZSP, the Chinese Picturesque was considered secondary to the primary influence of international design precedents from America and Germany. International design influences were represented in the four hybrid modern parks. Damon and Ruddick's work, as well as EDAW's work, were a result of foreign nationals trained abroad. Chinese native and Harvard-educated, Yu drew from his American experience and research on Haag's Gasworks Park in Seattle and Latz's work in Germany. HDI's designers were all educated in China and influenced by the ecological principles deployed in LWP and ZSP, as well as the craftsmanship of EDAW's work at Jinji Lake. In addition to the inspirations from their archival research (classical paintings and poetry) on West Lake, HDI's designers were inspired by their internet research of projects on professional websites at that time, especially works by Hargreaves Associates, Sasaki, SWA Group and Peter Walker Partners.

This book argues that these four hybrid modern parks represent something entirely different from the western 1990s postmodern world. This separate trajectory and distinction align with other scholars who have contested China's postmodernity. Many of them claim that modernity in terms of "modernism" did not evolve in China as it did in the West and therefore cannot have the postmodern experience (Zhang 2006; Wang 2003; Dirlik 2002; Strand 2000).

In this debate, China's cultural development never progressed to the postmodern moment experienced in the west and Japan. The claim is that the forty-year period of isolation caused by the Japanese occupation, famine, and isolation and suppression of cultural development (across various disciplines) in the Mao era arrested any development of China's modernity. This book's particular premise is careful not to imply "progress" and argues that China's path for cultural development will and has occurred on its own terms. Some argue that this represents discontinuity of cultural development and modernity, and others state that modernity was interrupted in the Mao period and the preceding period of warlords and civil strife. However, the case can be made for China's late 20th century hybrid modernity as a distinctive path when deeply investigating the four purpose-built parks in the secondary cities discussed in this book.

Following the trajectory informed by Castells (1989), Jameson (1998) and Anderson (1983), the book's premise argues that the late 20th century opening of China could not halt the media bombardment of new ideas and imagery, particularly with the multivalent change and global interconnection caused by the world wide web and digital age by the late 1990s and turn of the 21st century. International trends were bound to influence China's socio-cultural development, especially within the realm of landscape architecture and allied disciplines working in the built environment. The agency of the Chinese Picturesque garden tradition was crucial and unrelenting in the discipline of landscape architecture. The designers for the case study parks created didactic experiences where visitors and local citizenry could learn about local heritage and the natural environment, as well as new leisure and recreational experiences as symbols of the New Era. Taken together, these temporal, immersive learning and socio-cultural contexts created a trajectory for the resulting hybrid modern spatial forms in purpose-built parks in China's secondary cities. Table 5.5 summarizes hybrid modern design language in China's late 20th century and the design approaches deployed by the teams working on the four case study parks.

In China's modern park history, hybrid modernity represents a period when the four purpose-built parks represent a distinctive new design language for the post-Mao New Era. The four case study parks were built in China's secondary cities, places where Deng's reforms created competition among local (municipal, provincial government

Table 5.5 Summary of Hybrid Modern language for the four case study parks

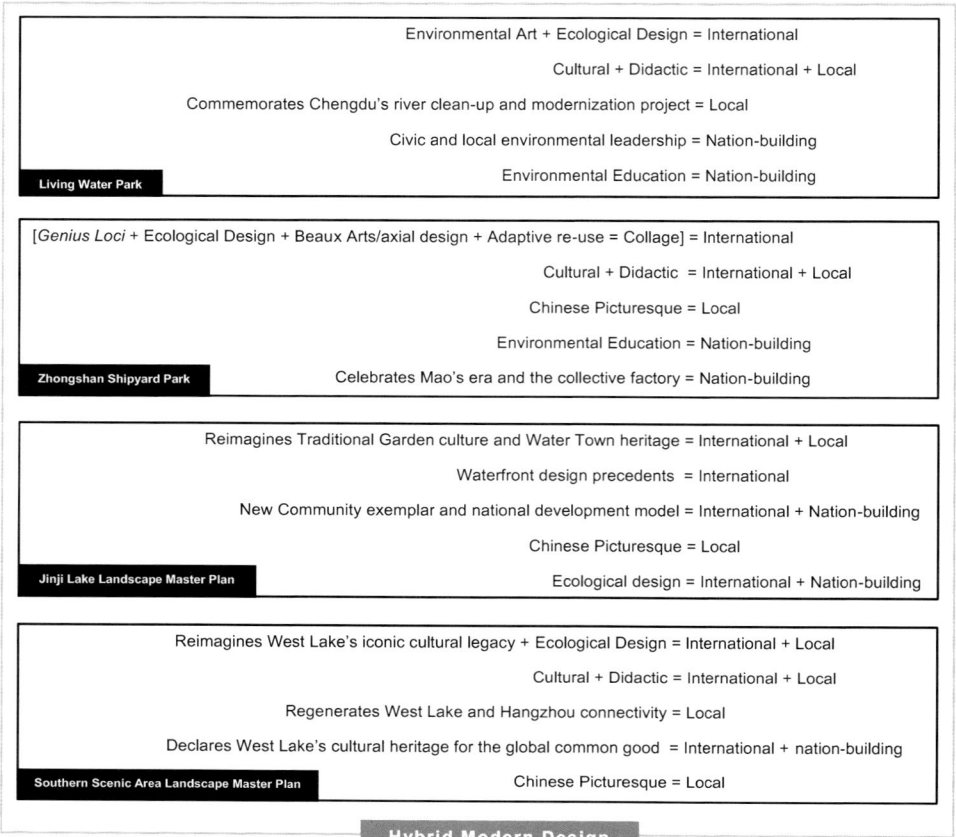

and CPC) leaders who engaged in innovation during the post-Mao urban experiment. As China awoke from Mao's period of scarcity, destruction, chaos and isolation, these new parks were considered "gifts" and new places where individual citizens in the emergent post-socialist society could feel liberated from a period of oppression and humiliation, and enjoy their new leisure time. Taken together with the history of parks since the 19th century, these four purpose-built parks are part of a larger mosaic narrative for understanding modernity in China (Figure 5.60).

The next chapter, "Transforming from Hybrid to Ecological Modernization in China's 21st century", examines and speculates on park design and the development of landscape architecture after China formally joined the World Trade Organization (WTO) in late 2001 – China's foray into global modernity as a major economic actor. It touches on urban development in the host cities of the four case studies and expands on broader trends in China's 21st century, especially in the aftermath of the 2008 Beijing Olympics. It reveals this period of development in landscape architecture as not unlike the post-1994 period in China's art world when works were created for international consumption. It brings to light China's top-down governance and Five-year Plan (FYP) national development model and ways China's "sponge city"

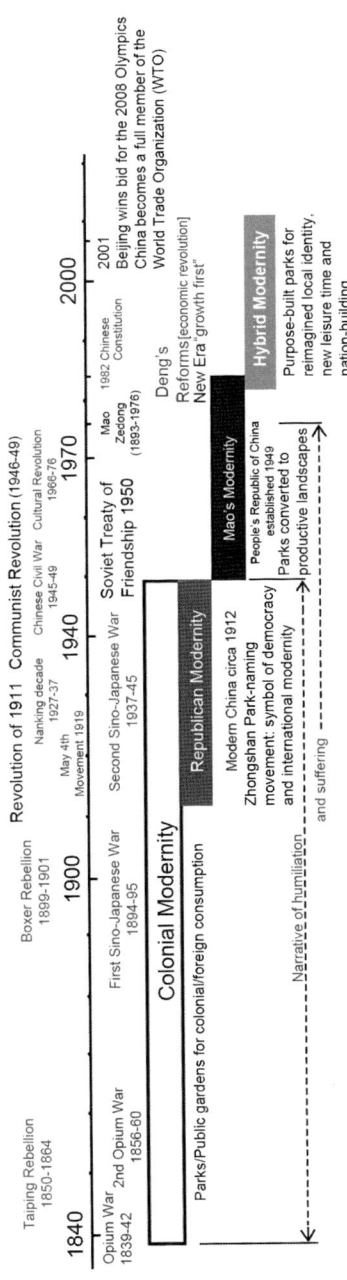

Figure 5.60 Timeline of park development suggests a mosaic narrative for understanding modernity in China, graphic by author

initiative expanded the lexicon of urban design, landscape architecture and ecological urbanism. It also brings to light China's 21st century green revolution triggered through China's top-down governance, and ways that President Xi Jinping's notion for developing "Beautiful China" is inextricably linked to environmental stewardship for both the national and international common good.

References

Amidon, J. (2001) *Radical Landscapes: Reinventing Outdoor Space*. New York, NY, Thames & Hudson.

Anderson, B. (1983) *Imagined Communities: Reflections on the Origin and Spread of Nationalism*. London, UK, Verso.

Baljon, L. (1992) *Designing Parks*. Amsterdam, Netherlands, Architectura and Natura Press.

Barmé, G. (2012) A Chronology of West Lake and Hangzhou, *China Heritage Quarterly*, 28. http://www.chinaheritagequarterly.org/features.php?searchterm=028_chrono.inc&issue =028

Bie, J., de Jong, M. & Derudder, B. (2015) Greater Pearl River Delta: Historical Evolution towards a Global City-Region, *Journal of Urban Technology*, 22(2), 103–123.

Castells, M. (1989) *The Informational city: Information Technology, Economic Restructuring, and the Urban-Regional Process*. Oxford, UK, Blackwell.

Chen, G. (2003) *Hangzhou*. Beijing, China, Foreign Language Press.

Clunas, C. (1997) *Art in China*. Oxford, UK, Oxford University Press.

Dirlik, A. (2002) Modernity as History; Post-Revolutionary China, Globalization and the Question of Modernity, *Social History*, 27(1), 16–39.

Ebrey, P. (2008) *Accumulating Culture: The Collections of Emperor Huizong*. Seattle, WA, University of Washington.

Fairbank, J. & Goldman, M. (2006) *China: A New History*. Cambridge, MA, Harvard University Press.

Fernsebner, S. (2002) *Material Modernities: China's Participation in World's Fairs and Expositions, 1876–1955*. Berkeley, CA, University of California Press.

Francis, M. (2001) A Case Study Method for Landscape Architecture, *Landscape Journal*, 20(1), 15–29.

Gao, J. Z. (2006) *The Communist Takeover of Hangzhou: The Transformation of City and Cadre, 1949–54*, Honolulu, HI, University of Hawaii Press.

Gonzalez, R. (1998) Waterworks Garden Submitted by Lorna Jordan Inc., Seattle, *Places*, 12(1), 17–18.

Huat, C. (2011) Singapore as Model: Planning Innovations, Knowledge Experts. In: *Worlding Cities: Asian Experiments and the Art of Being Global*, Roy, A. & Ong, A. (eds.). Western Sussex, UK, Blackwell Publishing Ltd.

Jameson, F., (ed.) (1998) *The Cultures of Globalization*, Durham, NC, Duke University Press.

Johnston, R. S. (1991) *Scholar Gardens of China: A Study and Analysis of the Spatial Design of the Chinese Private Garden*. Cambridge, UK, Cambridge University Press.

Keswick, M. (1978) *The Chinese Garden: History, Art and Architecture*. London, UK, Academy.

Lee, L. T. (1981) Deng Xiaoping's ASEAN Tour: A Perspective on Sino-Southeast Asian Relations, *Contemporary Southeast Asia*, 3(1), 58–75.

Lin, Z. (2015) The Building of a Chinese Model New Town: Case Study of the Suzhou Industrial Park, *ACSA Conference Proceedings*, Seattle, WA

Ma, L. J. C. & Wu, F. (2005) Restructuring the Chinese City: Diverse Processes and Reconstituted Spaces. In: *Restructuring the Chinese City: Changing Society, Economy and Space*, Ma, L. J. C. & Wu, F. (eds.). London, UK, Routledge.

Marme, M. (2018) From Suzhou to Shanghai: A Tale of Two Systems, *Journal of Chinese History*, 2(1), 79–107.

Mote, F. W. (1970) The City in Traditional Chinese Civilization. In: *Traditional China*, Liu, J. T. C. & Tu, W. (eds.). New York, NY, Prentice-Hall.

Padua, M. (2003) Industrial Strength, *Landscape Architecture*, 93(6), 76–85, 105–107.

Padua, M. (2004a) Future Scale, *Landscape Architecture*, 94(8), 106–115.

Padua, M. (2004b) Teaching the River, *Landscape Architecture*, 94(3), 100–109.

Padua, M. (2014) China: New Cultures and Changing Urban Cultures. In: *New Cultural Landscapes*, Roe, M. & Taylor, K. (eds.). London, UK, Routledge.

Perry, M. & Yeoh, C. (2000) Asia's Transborder Industrialization and Singapore's Overseas Industrial Parks, *Regional Studies*, 34(2), 199–206.

Prentice, H. (1998) *Suzhou: Shaping an Ancient City for the New China: An EDAW/Pei Workshop*. Berkeley, CA, Spacemaker Press.

Qian, Z. (2012) Post-Reform Urban Restructuring in China: The Case of Hangzhou 1990–2010, *Town Planning Review*, 83(4), 431–456.

Qin, B. (2015) City Profile: Chengdu, *Cities*, 43, 18–27.

Sage, S. (1992) *Sichuans Century under Qin: Ancient Sichuan and the Unification of China*. New York, NY, SUNY Press.

Schepelmann, P., Kemp, R. & Schneidewind, U. (2016) The Eco-Restructuring of the Ruhr District as an Example of a Managed Transition. In: *Handbook on Sustainability Transition and Sustainable Peace*, Brauch, H. G., Oswald Spring, Ú., Grin, J., & Scheffran, J. (eds.). New York, NY, Springer.

Shi, D. (2009) Recording Sixty-Years of Change at West Lake, *Chinese Landscape Architecture*, 4, April.

Skinner, G. W., (ed.) (1977) *The City in Late Imperial China*. Stanford, CA, Stanford University Press.

Steinhardt, N. (1999) *Chinese Imperial City Planning*, Honolulu, HI, University of Hawaii.

Stilgenbauer, J. (2005) Landschaftspark Duisburg Nord, *Places*, 17(3), 6–9.

Strand, D. (2000) A High Place Is No Better than a Low Place: The City in the Making of Modern China. In: *Becoming Chinese, Passages to Modernity and Beyond*, Yeh, W. (ed.). Berkeley, CA, University of California.

Wang, H. (2003) *China's New Order: Society, Politics, and Economy in Transition*, Cambridge, MA, Harvard University Press.

Wang, L. (1999) Tourism and Spatial Change 1911–1927. In: *Remaking the Chinese City*, Esherick, J. W. (ed.). Honolulu, HI, University of Hawaii.

Wang, Y. & Bramwell, W. (2012) Heritage Protection and Tourism Development Priorities in Hangzhou, China: A Political Economy and Governance Perspective, *Tourism Management*, 33(4), 988–998.

Way, T. (2015) *From Modern Space to Urban Ecological Design*. Seattle, WA, University of Washington Press.

Wright, A. F. (1977) The Cosmology of the Chinese City. In: *The City in Late Imperial China*, Skinner, G. W. (ed.). Stanford, CA, Stanford University Press.

Xu, Y. (2000) *The Chinese City in Space and Time: The Development of Urban Form in Suzhou*, Honolulu, HI, University of Hawaii Press.

Ye, X., Chen, C. & Eng, R. Y. (1990) Social and Economic History of Guangdong Province: State of the Field, *Late Imperial China*, 11(2), 102–115.

Yu, K. & Pang, W. (2002) *The Culture Being Ignored and the Beauty of Weeds*. Beijing, China, China Architecture & Building Press.

Zhao, S. (1993) Deng Xiaoping's Southern Tour: Elite Politics in Post-Tiananmen China, *Asian Survey*, 33(8), 739–756.

6 Transforming from "hybrid" to "ecological" modernization in China's 21st century

Twenty-first century trends in greening, purpose-built park design and landscape architecture in China are discussed in this chapter along with speculations on the future of the discipline. It also discusses China's top-down governance and its 20th and 21st century history of "reform" and "revolutionary" praxis. In this light, it introduces China's "green revolution" and ways this political context facilitated the growth of the discipline of landscape architecture. It theorizes the 21st century transformation and shift from Deng Xiaoping's "growth first" late 20th century "hybrid" modernization to "ecological" modernization when former President Hu Jintao's "people first" philosophy emerged circa 2007 in China. The chapter is organized to first review trends in the host cities of the four hybrid modern case study projects since their completion – Chengdu, Sichuan; Zhongshan, Guangdong; Suzhou, Jiangsu; and Hangzhou, Zhejiang, respectively, followed by a discussion on general trends influencing park-making, greening and landscape architecture throughout China. China's top-down web of governance including the national development policy instrument, the Five-Year Plan (FYP), is briefly unraveled with a discussion on the "Sponge City" urban initiative as an FYP exemplar. It also reviews the navigation of China's sponge city into the world lexicon of green design. The chapter closes with speculations on the growth of the discipline of landscape architecture through the lens of China's emergent "green" aesthetic and President Xi's pursuit for "Beautiful China, *Meili Zhongguo*" – a broader domestic and international imperative for environmental stewardship.

A few notes frame the chapter's discussion on landscape trends in China's 21st century. One critical discovery is the degree to which China's top-down system of governance creates a type of "ad hoc nation-building machine" that enabled the development of the landscape architecture discipline. Additionally, modern China's broader historical development evolved through a unique "reform" and "revolutionary" praxis. In this light, former President Hu Jintao's "putting people first, *yiren weiben*" philosophy circa 2007 is widely accepted as a reform to former paramount leader Deng Xiaoping's "growth first" policy, also referred to as China's economic revolution or radical reform to Mao's socialist era. Hu's people-first philosophy grounded his notions of the scientific development of China as a "harmonious society, *hexie shehui*" and "ecological civilization, *shengtai wenming*". Hu's leadership also spurred China's "green" revolution and shift to a circular economy, a type of sustainable development strategy that encompasses principles of reduction, re-use and recycling of materials, as well as energy efficiency and environmental protection. This laid the groundwork for President Xi's "Beautiful China, *Meili Zhongguo*", an environmentally driven concept that represents his legacy and commitment as a world leader.

Four case study parks and trends in the host cities

Living Water Park (LWP) in Chengdu, Sichuan Province is a purpose-built park that artfully demonstrates biomimetic ways to clean water. Completed in 1998, LWP was conceptualized by Betsy Damon, an American environmental artist, in collaboration with Margie Ruddick, an American landscape architect, with the final execution led by Chengdu's Landscape Bureau. The project commemorated Chengdu's larger modernization effort to clean the city's river corridor, initiated through a grassroots effort. LWP received numerous accolades and was on the requisite list of projects for official tours by government officials (Padua 2004b). It is also known as China's first ecological park, *shentai gongyuan*. Local experts and scientists determined the river's water pollution was caused by upstream non-point sources from rural areas, not solely from urban sources of domestic effluent and industrial waste discharge. Subsequently, several of these experts established the non-profit organization Chengdu Urban Rivers Research Association (CURA) in 2003.

CURA's mission focuses on the elimination of non-point source pollutants and protecting the river system – a continuation of the grassroots work that spurred the Funan Revitalization Project. CURA's work led to the 2005 pilot project in Anlong Village, Pidu District, located forty kilometers (twenty-five miles) northwest of central Chengdu. This rural ecological experiment involved CURA's engagement with villagers to create a resource-efficient community that reduced chemical use, farm pollution, and promoted organic agriculture and rural environmental education. By 2013, over 160 Anlong Village households were considered closed cycle eco-households (utilized bio digesters, gray water systems, constructed wetlands and composting toilets, among other resource-efficient practices) and participated in chemical-free farming, community-supported agriculture, and rural environmental education (Zhao 2013; Economy & Zhang 2015).

It's important to note that the rural patterns of villages around Chengdu reflect a unique and idyllic cultural and agrarian landscape form known as *"linpan"* (Figure 6.1). This settlement pattern is defined by a patchwork of single farmhouses nested among trees surrounded by productive landscapes and irrigation trenches with its water sourced from the nearby ancient Dujiangyan Irrigation system, a designated UNESCO World Heritage site (Li et al. 2019). CURA efforts sought to preserve this cultural landscape with their efforts at the Anlong Village, a *linpan* settlement, and as a successful sustainable rural experiment they expanded the CURA founders' previous transformative experience at Chengdu's LWP and commitment to "living water". This village aligns with China's nation-building efforts to integrate the development of urban and rural settlements.

Chengdu, the capital city of Sichuan Province, grew in its national and international significance since the completion of LWP in 1998. The city was identified in China's 2000 "go west" development strategy where the central government provided support for infrastructure construction, foreign direct investment, environmental protection and education. Investments in Chengdu's special economic zones from over 200 Fortune 500 firms have been recorded and the city claims the positive influence of foreign expatriate residents. In this context and China's entrepreneurial and competitive nature of local governance, Chengdu's mayors launched several political campaigns: "garden city", "rural-city unification" and more recently "park city, *gongyuan chengsi*" (Kuo 2019).

Figure 6.1 Linpan development, photograph provided by Chengdu, published with permission

The river and environmental sustainability, along with healthy lifestyles, remain important to local officials. Since LWP's completion, Chengdu has embarked on a vast program of public greening including the construction of several urban parks, suburban wetland parks, green belts and greenways, and environmental protection zones. This translated into certain indices (greening and forest coverage among others) that deemed it a National Forest City (NFC) in 2007. The NFC program was initiated in 2004 and refers to a city with an urban ecological system and forest cover that meets the standards set by State Forestry and Grassland Administration, a government unit managed by China's Ministry of Natural Resources. Chengdu also qualified as a National Ecological Garden City, an upgrade to their previous status as a National Garden City. This previous Ministry of Housing and Rural-Urban Development program was largely based on beautification (Shi et al. 2018). Chengdu was also deemed a "National Civilizing City, *Quanguo wenming chengshi*", by central government's Spiritual Civilization Commission, an award highly coveted by local politicians that reflects and validates their commitment to environmental sustainability, healthy living and a people-centered place; and in 2014 the Asia Development Bank ranked Chengdu as China's top livable city. Positive international assessments and metrics assist local officials' political advancement.

A national spatio-temporal study of public parks in China indicated Chengdu's production of eighty-one public parks between 1981 and 2014; and it noted 2000 as a milestone year when park development started accelerating throughout China (Wang & Liu 2017). This validates the growing significance of landscape architecture in China since the turn of the 21st century. Chengdu's vast greening efforts since LWP was completed in 1998 included new and renovated wetland parks that were intended to meet the 2015 State Council directive for "Sponge City Construction".

Table 6.1 State Council 2014 city classifications by population size, based on Xinhua (2014) and Gipouloux & Li (2015)

Tier	City Classification	Population Size
1	Super-large (megacity)	+ 10 million
2	Very-large	between 5 and 10 million
3	Large	between 1 and 5 million
	Type I	between 3 and 5 million
	Type II	between 1 and 3 million
4	Medium	between 500,000 and 1 million
5	Small	less than 500,000
	Type I	between 200,000 and 500,000
	Type II	less than 200,000

Chengdu and all cities in China are required to update their 20-year development plans for central government approval. The urban population was around 4 million in 1998 and Chengdu's larger administrative area had grown to a population over 18 million, qualifying it as a Tier 1 "super large-sized city" in China's 2014 classification of cities according to population (Gipouloux & Li 2015) (Table 6.1). Chengdu's approved 2016–2035 revised plan called for the Tianfu Greenway, a vast multi-functional green open space system that commenced construction in 2017. When completed in 2035, the project will traverse 20,000 kilometers (6213 miles) and create 1.45 million hectares (3.6 million acres) of continuous "green" space. This greenway project, a landscape system, is intended to link public parks, wetlands and forest parks, cultural heritage sites, nature reserves, wildlife habitats and ecological protection zones, and more significantly, provide places for people to socialize and spend their leisure time. It is also intended to encourage "active living" where residents and visitors walk, ride bicycles and exercise. Furthermore, Tianfu Greenway provides assembly areas for cultural events, sports, tourism, emergency shelter, and rural regeneration opportunities for small gardens and micro-green agricultural lands.

The second case study, Zhonghsan Shipyard Yard Park (ZSP) in Zhongshan, Guangdong Province, was completed in 2001 and designed by Yu Kongjian and his firm, Turenscape, China's first private (non-state-owned enterprise) landscape architecture company. ZSP is the first of its kind in China and involved the adaptive re-use of a brownfield site, an existing industrial shipyard collective work unit (Padua 2003). Turenscape's design was organized around celebrating the "spirit of place" and ecological design. It also symbolized cultural awakening with its transformation from an industrial site into a new type of park for the post-Mao era and celebrates the site's history as a Mao era communal factory.

ZSP's spatial forms were informed by the site's industrial past, the city's flood control needs and natural resources. As a former shipbuilding site, it was located along a river with several mature trees along its riverfront edge. To deal with the city's needs to widen the river, Turenscape created an island along its riverfront edge for largely ecological purposes. Considered heroic landscape architecture at the time, this so-called ecological island preserved the mature trees and the shipyard's historic boundary while meeting the city's need to widen the river for flood control. ZSP was designed as an immersive cultural and educational experience where visitors could learn about Mao's industrial age and local history of shipyard building, the site's natural environment and contemplate the Cultural Revolution. According to Yu, "it's a small site with a big story" (Padua 2003). It was the first designed landscape that publicly acknowledged the Cultural Revolution – a ten-year period that most people in China, especially the central government, wanted to forget. ZSP's design excellence received prestigious international awards and launched Yu's career in China and internationally.

According to China's classification of cities based on population, Zhonghsan, Guangdong Province, is a "super city" (between 1 and 5 million population). In China's 2000 census (conducted every ten years), Zhongshan's population was recorded at 2.36 million with its current population estimated at 3.23 million. Its local reputation is tied to its namesake, Sun Yat-sen, and its international reputation is linked to the Pearl River Delta region's earlier standing as the "workshop of the world". To maintain its status as a National Garden City, awarded in the late 1990s, Zhonghsan continued its greening and park-making efforts including the establishment of various new parks, and imposed requirements of greening and public parks for any new development projects, whether residential, commercial or high technology. Like Chengdu, Zhongshan was deemed a National Civilized City. China's Ministry of Environmental Protection (restructured and expanded in 2018 to the Ministry of Ecology and Environmental Protection) designated Zhongshan as a National Model for Environmental Protection and National Ecological City. Adding to their "green" local identity, Zhongshan was designated a National Forest City (NFC) in 2018 by China's State Forestry and Grassland Administration. Greening, park-building, ecological restoration, and environmental protection are seen as foundational for Xi's Beautiful China.

Of critical significance for Zhongshan was China's National Development and Reform Commission's (NDRC) recent 18 February 2019 launch of the "Guangdong-Hong Kong-Macao Greater Bay Area Outline Development Plan (GBA-ODP)", a four-party cooperation agreement involving the NDRC, Guangdong Province, Hong Kong Special Administrative Region (HKSAR) and Macao Special Administrative Region (MSAR). The GBA-ODP is a near-term (2019–2022) and long term (up to 2035) development strategy for HKSAR, MSAR and the Pearl River Delta (PRD) cities: Guangzhou, Shenzhen, Zhuhai, Foshan, Huizhou, Dongguan, Zhongshan, Jiangmen and Zhaoqing. It follows within the spirit of the 19th National CPC Congress (held every five years) circa October 2017, and aligns with the National New-type Urbanization Plan (NUP) 2014–2020 (Figure 6.2) where the PRD is a designated mega-city cluster or urban agglomeration. The NUP organizes China's development of mega-city clusters along "*liang heng san zong*, two horizontal, three vertical axes" (Fang 2015). GBA-ODP's ambitious goal is to create a dynamic hub of innovation, financial services, shipping and related businesses that rival California's Silicon Valley (Cheung & He 2019). It's important to note that economic and physical interconnectivity among the various PRD cities was a typical practice before the GBA-ODP launch. For example, much of Shenzhen's hyper-rapid conversion from a fishing village to a skyscraper city was due to Hong Kong capital and a similar pattern exists with Macao's business investments in Zhuhai. Zhongshan is also connected to Hong Kong as a haven for "second" homes or weekend vacations, especially given it is less than a one-hour ferry ride from Hong Kong. Physical and economic connectivity is likely to expand Zhongshan's PRD presence given bridge construction scheduled for a 2021 completion, an infrastructure project that connects it to Shenzhen within thirty minutes by car. Despite the GBA-ODP launch, according to the city's website,

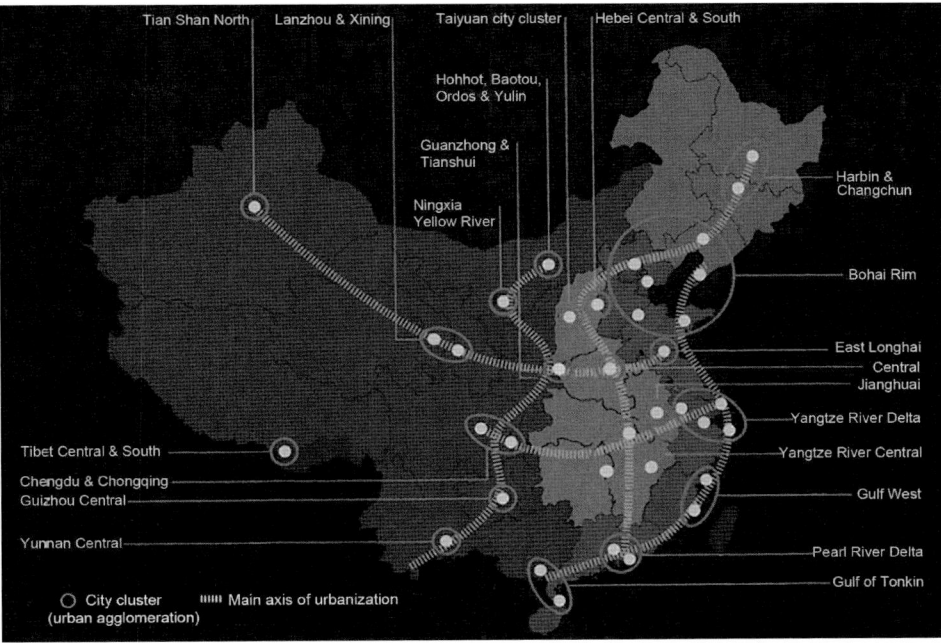

Figure 6.2 National New-type Urbanization Plan 2014–2020, based on Fang (2015), graphic by X. Liu

Zhongshan's identity revolves around its revolutionary namesake, "green" and livable image, and its claim as the ancestral home to over 800,000 Chinese living abroad.

The third case study, Jinji Lake Landscape Master Plan (JLLMP), in Suzhou, Jiangsu Province, was completed circa 2002–2003 and designed by one of America's largest landscape architecture and environmental design firms, EDAW (acquired by AECOM circa 2005) in a joint venture with the city of Suzhou's Landscape Bureau. The project's public open space elements were considered critical public infrastructure for the Sino-Singapore cooperation zone located in Suzhou's eastern expansion area, known as Suzhou Industrial Park (SIP), a model town or "new community" for domestic and foreign professionals in the financial and high technology sectors with related housing. The JLLMP established a public open space system with designed landscapes that linked the "new community" in the eight surrounding districts with Jinji Lake (Padua 2004a). Largely culture-based with elements of ecological design, the JLLMP honored the local water town heritage and Suzhou's classical garden culture. It also incorporated an international waterfront design, particularly given the client's objective to house international professionals. AECOM's Hong Kong-based Vice President and Director of Landscape Architecture, David Jung (former EDAW principal designer), noted his satisfaction with the execution of "slow" landscape architecture around the lakefront with its continuous bicycle and pedestrian path system and community open space linkages, and its conformance to their original design intentions (10 June 2019 interview). SIP has since been completed, built out with over 1 million people working and living in a national new town model – a green environment based on a Singapore urban planning, public administration and economic model with foreign direct investment (Li 2019).

During SIP's development, historic Suzhou experienced the renaissance of Pingjiang Canal historic district with the 2006 completion of the Suzhou Museum. This museum and garden complex was designed by master American architect, I. M. Pei (1917–2019). Architectural critics noted Pei's museum design was largely a personal journey, given he spent time in Suzhou during his youth; and it was a project that didn't break any new ground in China (Barboza 2006; Ivy 2007). However, the museum's gardens and courtyards represent an elegant reimagining of Suzhou's classical garden tradition. In this light, a few points are important for understanding the 21st century evolution of designed landscapes in China.

Pei noted the museum's garden spaces were largely vestigial spaces, given his focus on the building design. He recognized gardens were typically more prominent as organizing features in the Suzhou tradition of designing private gardens in a residential building complex (Pei interview, 19 May 2009). However, during visits to the Suzhou Museum, the author observed visitors were immediately drawn to the museum's waterscape, a dominant outdoor feature visible to visitors entering the museum building complex. Visitors are able to physically access an outdoor platform and immerse themselves in Pei's reimagined mountain-water (*shan-shui*) courtyard garden. From this vantage point, the stationary garden view is somewhat reminiscent of the Ryoanji Zen temple courtyard garden in Kyoto, Japan. In that context, visitors experience a panoramic view of the temple garden while sitting or standing. Instead of raked sand, the foreground view at the Suzhou Museum is dominated by a still reflecting pool with a level stone bridge traversing it. A few of the Suzhou gardens have a similar experience. For example, at the Yi Pu garden, visitors first experience a large waterscape scene. Most of Suzhou's Chinese Picturesque scholar gardens usually entail an experiential progression of moving through a series of scenes in courtyards and gardens via "twists and turns" in the path system.

In reviews of the Suzhou Museum, little is mentioned about Pei's inspiration for the garden design. The author discovered there was more to his public garden design strategy than the "fill-in", the vestigial voids left over from his building design. Pei's goal was to introduce his contemporary interpretation of the local Chinese Picturesque garden tradition that included vernacular architecture to Chinese politicians and future designers for critical study. In this light, Pei's project can be construed as hybrid modern – appropriating from the local Chinese Picturesque garden tradition, vernacular architecture in China's canal towns, international approaches and to an extent, nation-building.

Pei's building footprint and site plan incorporated a series of courtyards with varying sizes and themes, and occupies the former grounds of the adjacent historic Prince Zhong cultural building complex. Pei's garden design took advantage of a whitewashed stucco wall shared with the adjacent complex and used it as a contrasting backdrop for the dark stones that comprised an abstract mountain-scape (Figure 6.3). According to staff in Pei's New York office, the cut stones were sourced from a quarry at Mt. Tai, *Taishan*, in Shandong Province. The waterscape in the museum's main courtyard is elegantly executed and Pei noted his inspiration for the garden's mountainscape was a painting by Mi Fu, a Song dynasty artist whose work represents the "misty" style genre. Pei considered this particular garden scene as an artistic experiment where he "painted with rocks" – reinforcing his understanding of the Suzhou

Figure 6.3 Pei's garden represents his experiment to "paint with rocks" at the Suzhou Museum, photograph by author

classical garden as a three-dimensional expression of Chinese classical arts. He noted the time working onsite with craftsmen on finishes and the final spatial arrangement of the abstract mountain-scape was a period he cherished tremendously. The selection of plants throughout the site mimicked the Suzhou scholar-official garden with individual tree plantings selected at local nurseries and grafts of a 500-year-old wisteria vine from the adjacent historic Prince Zhong complex. Pei's hybrid modern garden is referential to the Suzhou classical garden tradition and was created for both national and international consumption.

After Pei's Suzhou Museum was completed and SIP's various phases were completed around Jinji Lake, Suzhou, along with Chengdu and Hangzhou, they were included in the 2010 National Low Carbon City pilot program launched by China's National Development Reform Commission (Wang et al. 2015). Urban green space was a critical parameter for this national-level pilot program. In 2011, Suzhou was also deemed a National Ecological Garden City by the Ministry of Housing and Urban-Rural Development, one of the inaugural twelve cities selected from 184 National Garden Cities (Zhou et al. 2012). Later in 2015, SIP's administration received central government approval for the SIP 2.0 plan, the next generation development strategy as China's first experimental zone on "opening-up and innovation", and was re-branded as "innovation paradise".

SIP 2.0 builds from SIP 1.0 sustainable development and environmental exemplar in China with its "green" facilities for waste and pollution treatment. Its livable environment, largely created through the implementation of the Jinji Lake Landscape Master Plan, contributed to its status as an exemplar for sustainable development. Along with Hangzhou, Shanghai and twenty-three other cities, Suzhou is part of the Yangtze River Delta (YRD) mega-city cluster, one of nineteen clusters designated in

China's NUP (Figure 6.2). Chengdu, along with Chongqing, comprises another of the nineteen mega-city clusters located at the western terminus of one of the two horizontal axes in China's NUP; and as noted earlier, Zhongshan is part of the PRD mega-city cluster in southern China.

The fourth case study, West Lake Southern Scenic Area Master Plan (SSAMP) in Hangzhou, Zhejiang Province was completed in 2002 and designed by the Hangzhou Landscape Architecture Design Institute (HDI) established in 1952 (Padua 2014). HDI's project focused on the design of a continuous lakefront path and system of public open spaces along West Lake's urban edge and Nanshan Street. It reimagined West Lake's classical mountain-water, *shan-shui*, iconic landscape imagery, a key cultural component of China's national landscape heritage and local urban history. The project celebrated West Lake's link to Hangzhou's urban heritage: royal capital city in the earlier Five Dynasties and Ten Kingdoms period; China's temporary capital city in the Song dynasty; and modern China's 20th century city-making with a new lakefront system of public parks along the alignment of the demolished historic city walls. Late 1990s new Hangzhou leadership understood the importance of this legacy and expanded West Lake's appreciation to the level of international cultural consumption, especially for the global common good.

HDI's hybrid modern design considered the area's landscape legacy, urban heritage and ecological design including West Lake folklore and mythology, the lake's physical, cultural and historical relationship to Hangzhou, as well as the lake's water quality and natural systems. The project covers an area over 300 hectares (740+ acres) and is spatially organized into nine various lakefront components and three key public open space projects along Nanshan. HDI's work contributed to the 2011 designation of "West Lake Cultural Landscape" as a World Heritage site by the United Nations Educational, Scientific and Cultural Organization (UNESCO). In China, attaining a UNESCO designation is considered a major nation-building achievement. Unlike the other three case study projects, West Lake is part of a National-level Scenic Area, *Guojiaji Fengjing Minshengqu*, the equivalent of a national park in western nations, and listed in China's 1982 inaugural list of forty-four national-level scenic areas.

As the capital of Zhejiang Province, local and provincial leaders have been active in their efforts to maintain Hangzhou's status as a National Ecological Garden City and as one of the inaugural cities for the National Green Low Carbon City pilot program. Like Chengdu and Suzhou, Hangzhou was designated a National Civilized City, a highly coveted award for local officials who demonstrate their city is engaged in developing as a harmonious society. Like the other host cities of the case study parks, Hangzhou's population grew from 3.6 million in 2000 to around 5.2 million according to the 2010 census and was estimated at 7.2 million in 2018. According to the city's website, Hangzhou is also considered a "smart city" and self-made "Silicon Valley", given it's the birthplace of Alibaba (e-commerce and retail equivalent to Amazon) along with other Chinese technology start-ups and well-known universities in the area.

To deal with population expansion and sustain their status as a Low Carbon City, Ecological Garden City and Civilized City, Hangzhou engaged in various greening efforts since SSAMP's completion. This included another major HDI work known as Xixi Wetland Park located twelve kilometers (7.5 miles) northwest of West Lake. Completed in 2005, it was later designated China's first national-level wetland park. It occupies eleven square kilometers (4.5 square miles) and integrates wetland farming practices, the preservation of the site's cultural heritage and establishes a natural wetland conservation zone. It exemplifies Hangzhou's water town and agrarian heritage, as well as the fact it was a destination for the literati during China's classical imperial era. HDI's unique design created an immersive and didactic experience for visitors to understand the area's cultural and agrarian heritage, as well as appreciate wetlands as a natural resource.

According to Hangzhou's website, their updated 20-year plan notes urban greening goals: the city's urban core as a "green ecological space" with "eco-belts" integrating and linking the urban core with three designated sub-areas or development "points". It also notes Hangzhou's long-term vision encapsulated by "mountain, lake, city, river, farmland, sea and stream" through the implementation of "four parks (suburban forest parks), multiple green areas (water conservation areas, wetland and scenic areas, and environmental protection zones) and multiple green corridors (waterfronts and transportation corridors)". To date, the city has implemented a network of greenways along canals and the Qiantang River and greenbelts along roads that link various types of public open spaces. HDI's West Lake project remains relevant and the most significant for Hangzhou and the central government in terms of local, domestic and global appreciation for the common good. It also contributes to China's nation-building goal for ecological civilization.

Trends in landscape architecture since the hybrid modern case study parks were completed indicate expanded open space networks and landscape systems or "green lungs" that connect urban areas with suburban and rural areas, and as places for everyday leisure and active living, and environmental restoration and protection with hints of agrarian landscapes for rural regeneration opportunities. It also demonstrates the importance of greening and public park-making throughout 21st century China. In this light, the discipline of landscape architecture has matured since Deng's reforms and accelerated since the turn of the 21st century. Environmentalism, restoring natural systems and conserving cultural and natural resources, as well as agrarian landscapes, appear to be driving 21st century trends in China's discipline of landscape architecture. Another trend addresses the growing social need to encourage active living and healthy communities, especially with China's goal to develop as a harmonious society and ecological civilization. These trends align with Xi's concept for "Beautiful China, *Meili Zhongguo*" announced at the 19th CPC National Congress (held every five years) in October 2017. Xi frames this nation-building concept positively, "building a better environment; protecting the ecosystem; and promoting green development improves national well-being" (Xi 2019).

Top-down China + Five-Year Plan = green revolution and ecological modernization

The mounting evidence for environmental neglect and deterioration of China's natural resources in the face of unregulated hyper-rapid urbanization rose to an international level as Beijing was preparing for the 2008 Olympics. The 2005 report by the Chinese Academy on Environmental Planning highlighted serious public health hazards created by China's polluted air and water; and around the same time the US Environmental Protection Agency reported that the source of 25% of particulate matter in the skies above Los Angeles, California, on certain days, were attributable to China (Watts 2015). In an address to CPC leaders in February 2005, President Hu raised the importance of shifting from aggressive "growth first, repair later" practices to balanced development through the making of a "harmonious society, *hexie she-hui*". Later in October 2007, the 17th National CPC Congress endorsed Hu's thinking on the scientific development of China as a harmonious society and "ecological civilization, *shengtai wenming*", – prioritizing human-centered development or "people-first" through energy efficiency, resource conservation, and environmental protection. In this light, China's leaders acknowledged the relationship between the nation's tremendous environmental destruction and the consequential public health crisis, and shifted priorities away from aggressive economic development towards remediating pollution and restoring natural resources. Hu's "people-first" philosophy triggered China's green revolution and informed the 11th Five-Year Plan 2006–2010, a critical top-down policy examined below.

Understanding China's top-down governance reveals ways Hu's 2005 speech on harmonious society influenced the national-level Five-Year Plan (FYP) development policy and triggered the green revolution. China politically functions through a single-party (Communist Party of China – CPC) system of centralized government with four branches: executive, legislative, judicial and military. The nation largely operates as a government hierarchy with four core levels: central; provincial-level jurisdictions; cities or counties; and townships and villages. Nation-building policies and priorities are set by the FYP, a policy instrument imported from the Soviet Union during the Mao period. However, it has evolved significantly from its initial static use with a shift occurring in the 9th FYP (1996–2000) when it became adaptive from one FYP to the next (Heilman & Melton 2013). During the same FYP time period, a parallel process of formulating and launching "Guidelines" and "Outlines" ensues among cabinet-level ministries and administrative units, provincial-level and local governments. Additionally, self-assessment and evaluative processes are formulated. This flurry of activities creates a culture of chaos, competition and uneven activities among the various ministries and administrative units.

A brief review of the FYP process and previous FYPs suggests China's political complex enables rapid change in a chaotic and ad hoc setting. It illustrates China's capacity to implement projects of any size or scale in a time frame that is difficult to accomplish in other places. It brings to light the shift from the late 20th century "growth first" period of hybrid modernization to "ecological modernization" and Hu's "people-first" approach by the end of China's first decade of the 21st century. In this theoretical context, ecological modernization draws from social theory as a reflexive phenomenon that is manifested in transformations in the government apparatus to overcome that nation's environmental crisis (Mol 1992; Giddens 1998; Harvey 1996). China's top-down approach, especially with their President as both head of the executive branch and CPC, and the nation's acknowledgment of irreparable environmental impairment created by unregulated growth, triggered the green revolution and enabled the nation's shift from hybrid to ecological modernization.

Formulating and implementing the FYP is a complex continuous cycle of information-gathering, policy analysis and formulation, evaluation and review. The FYP process is guided by three key principles: 1) it sets out policy directives for local governments to comply with; 2) it connects accomplishments with local achievements and assessments for career advancement; and 3) as a central planning policy apparatus, it spurs market activities (Heilman & Melton 2013). Each FYP is initiated two years prior to its launch following the mid-term review of the FYP in place. In this rapid process, FYP priorities and programs are subsequently carried out in a series of sub-plans through China's inter-locking hierarchical political web of four levels of government. It's important to note this flurry of FYP activities combined with the various government operations, and CPC leadership changes lay at the foundation of China's tremendous transformation and urban-rural experiment since Deng's reforms. A Chinese colleague describes this process as "constantly running to catch the fast-moving train within a shifting bureaucracy of technocrats".

Hu's human-centered development approach framed the nation-building policy directives in the 11th FYP (2006–2010) and formalized the "green revolution". Hu appropriated from Chinese philosophy on "harmony with nature" or man's "unity with nature" with a socio-economic vision that placed people's welfare ahead of economic development. Hu's vision of harmony was comprehensive and noted "five pairs of coordination, *wuge tongchou*" were critical to address China's social disparities: rural versus urban, coastal versus central and western, economic versus social, human versus nature, and domestic development versus openness to the world (Fan 2006). Hu's harmonious society is imbedded in "ecological civilization, *hexie shehui*", a concept enshrined in China's Constitution circa 2012 that shifts away from "grow first, clean up later" to nation-building activities that respect ecological carrying capacities. It is within this light that the world's largest urban experiment in China shifted to ecological modernization and created the crucible for the discipline of landscape architecture to flourish.

Highlights of "green" or environment-based and resource-efficient directives from past and current FYPs illustrate the reflexive process and progression of "green" thinking. It also illuminates the meaning of China's green revolution. The 9th FYP 1996–2000 includes a guideline for China's reliance on science and technology as a strategy for sustainable development, as well as "strengthening environmental and ecological protection and rationally developing and using natural resources". The 10th FYP 2001–2005 called for greening and increasing forest coverage in cities, as well as activities to improve air and water pollution; the 11th FYP 2006–2010 aligned with Hu's "people-first" approach and called for the reduction of energy consumption and a circular economy, the first time a nation anywhere in the world formalized a national development strategy based on closed-loop cyclical or eco-industrial processes, where waste products are converted into resources; the 12th FYP 2011–2015 highlights environmental protection as a "pillar industry" along with information technology and biotechnology, and called for low carbon technologies across various sectors; and the 13th FYP 2016–2020 was structured around five principles: innovation, coordination, green development, opening-up and sharing, along with a focus on environmental protection, development of low carbon technologies and improving people's livelihood and well-being. The notion of a "green" economy and environmental priorities evolved over twenty years through China's FYP process with the green revolution formally triggered by the circular economy in the 11th FYP and bolstered in subsequent FYPs.

It's important to reiterate that targets and metrics are further defined in FYPs or sub-plans by China's various State Council ministries and administrative units. Various initiatives were launched by various ministries and administrative units to meet these targets, including the Low Carbon pilot program, National Eco-city, National Ecological Garden City, an updated version of the 1992 National Garden City program, and the Sponge City national "green-building" initiative examined later in the chapter (Zhou et al. 2012; Wu & Padua 2019). These "green" programs individually have different conceptual parameters but broadly consider environmental degradation, climate change, the move away from fossil fuels consumption towards energy efficiency and renewable energy, especially in light of ecological civilization (Wang et. al 2015; Shi et al. 2018; Wu & Padua 2019). It's important to note that similar types of "green" development activities are initiated at provincial and local levels with some in collaboration with international non-governmental organizations (NGOs) or industries.

The consequences of a rare December 2015 meeting of the National Urban Work Conference further illustrate ways China's top-down governance influences landscape architecture and city-making. The last conference took place in 1978 when 18% of China's population lived in cities and this statistic was over 56% in 2015 (Ma 2015). Within a few months of the December 2015 meeting and prior to the 13th FYP 2016–2020 official launch, China's State Council announced the "Green and Smart Urban Development Guidelines". These national guidelines set basic principles, key tasks and targets so that cities would be "orderly constructed, properly developed, and efficiently operated" (Xinhua 2016). The overarching green directive limited cities from growing beyond the capacity of their natural resources. Other general policy directives related to landscape architecture and city-making included retaining distinctive landscape and cultural identities of cities; preserving historical and cultural heritage of cities; and building cities with beautiful natural landscapes. Some specific guidelines called for cities to develop more green belts and urban parks with no entry fees within close proximity to residents for their use, and intensify efforts to improve the environment (Xinhua 2016). Some qualitative and punitive comments regarding China's built environment threaded through the 2016 guidelines; for example, "bizarre architecture that is not economical, functional, aesthetically pleasing or environmentally friendly will be forbidden" (Zheng 2016). Some professionals working on projects in China, as well as theorists and critics, see the creative potential of the 2016 guidelines to transform the nation's bland and somewhat monotonous urban aesthetic, or China occupied by a "thousand cities with the same face". At the same time, these guidelines suggest a shift to cultural conservatism, a much deeper discussion that will be lightly touched on later in the chapter.

China's Sponge City program and contribution to international "green" urbanism

A review of China's Sponge City pilot program (SCPP) further illustrates China's top-down process and growing impact of landscape architecture. It demonstrates the urgency of China's response to dealing with the consequences of climate change and hyper-urbanization, as well as Xi's sense of domestic and global environmental stewardship. China's SCPP was largely a response to urban flooding created by sub-standard stormwater infrastructure, rapid urbanization and disrupted natural hydrological systems, and extreme weather events caused by climate change. It is considered an adaptive and resilient strategy with broader human-oriented and environmental goals for water conservation and flood control, and water scarcity.

In China, the so-called "sponge" concept is considered a biomimetic process and a form of green and blue stormwater infrastructure that slows, spreads, absorbs, filters and stores surface water run-off for later re-use. Aspects of this method have already been applied in Seattle, Washington; Singapore; and Sydney, Australia, and classified differently in each country, as well as in this book's four case study projects. For example, "low impact development (LID)" originated in Maryland, USA circa early 1990s, and has also been deployed as "Sustainable Urban Drainage System" (SUDS) and Blue-Green Cities (BGCs) in the United Kingdom (UK), Water-Sensitive Urban Design (WSUD) in Australia and Climate Proof Cities (CPCs) in the Netherlands (Workman 2017; Wu & Padua 2019). In China, the SCPP is intended to supplement existing conventional stormwater infrastructure that captures surface water through a subterranean conveyance system of inlets and pipes.

China's central government acknowledged public infrastructure could not keep up with hyper-urbanization and natural systems were harmed in this process. As a type of resilient design strategy, the sponge system mitigates flood waters from storm surges during extreme weather events, and deals with water scarcity by storing cleansed water for later re-use. Ideally in this system, surface water flows are slowed, treated and filtered through natural systems, and ideally captured, harvested and re-used. In terms of application, generally, paved areas in Chinese cities are retrofitted with a combination of vegetation, porous paving, water harvesting vessels and a re-distribution network. It also considers green roofs and innovative ways to capture rainwater. The brief narrative below notes ways that China's "sponge" concept emerged nationally and contributed to the world lexicon for "green" urban design and ecological urbanism, especially landscape architecture and allied disciplines.

The dramatic increase of economic losses from China's urban flooding between 1998 and 2014 brought to light the inter-relationships of hyper-urbanization and sub-standard stormwater infrastructure and related issues: deforestation, soil erosion, loss of natural resources and wildlife habitats; polluted surface water discharging into waterways and continual harm to natural water sources; and climate change and extreme weather events. Loss of life during flood events was also considered devastating, as well as major disruptions in everyday life in over 200 Chinese cities during the 2013 flood season and expectations for ongoing extreme weather (Wang & Guo 2010). The concept of sponge cities emerged in official meetings in China. Within a year of the floods, President Xi raised the importance of sponge cities and related water quality during an official speech in December 2013; and like all official presidential speeches, it was broadcasted, transcribed and disseminated through a variety of media communications. Additionally, in late 2012 Turenscape received an international design award (ASLA website) for "A Green Sponge for a Water-resilient City: Qunli Stormwater Park", completed in 2009 and located in Harbin, Heilongjiang Province. It's important to note that green infrastructure, LID stormwater best management practices, and cleansing water through biomimetic processes and ecological design were applied in the book's four case study parks completed between 1998 and 2002. This signals landscape architecture as the vanguard discipline involved in the development of China's sponge cities.

China's "sponge" design concept spread nationally and around the world soon after Xi's 2013 speech when he emphasized the transformation of China's urban areas into sponge-like cities that naturally accumulate, filter and purify rainwater (Leach 2016). Within two years, implementation of China's SCPP commenced and involved a multi-agency cooperative effort that led to the 2015 "Guiding Opinions of the General Office of the State Council on Advancing the Construction of Sponge Cities", formal launch of the SCPP and eventual selection of thirty official sponge cities out of over 500 applications by 2016 (Jia et al. 2017; Wu & Padua 2019). The SCPP came into play during the transition between China's 12th FYP and 13th FYP. As green public infrastructure, it was understood as contributing to economic growth, environmental protection and natural resource restoration, people's livelihood, as well as overall green innovation and nation-building. In this context, the SCPP aligned with the FYP's major indicators and expected to assist government leaders from all levels during their assessment period.

Since Xi's 2013 speech, the "sponge" term became visible in the international design lexicon in ecological or green urbanism. Cases of two international sponge city concepts were revealed in 2017 and another more recently. In August 2017, Berlin's local government released their updated climate change-oriented city plan (*Stadtentwicklungsplan /SteP Klima*), entitled "*SteP Klima KONKRET*", with a resilient goal to transform Berlin into a sponge city, or "*stadtschwamm*" in German (Berlin Journal 2017). Berlin's updated plan calls for a variety of sponge-related infrastructure including retrofitting old buildings and building new buildings with livable green roofs; establishing urban wetlands; and integrating permeable surfaces that absorb and store water during heavy rainfalls; and planting more trees for climate control. In northern California's San Francisco Bay region, another pro-active climate change project entitled "Resilient By Design (RBD) Bay Area Challenge" was launched in 2017: a multi-agency effort that involved nine multi-disciplinary teams working with nine different communities to investigate the protection and restoration of the bay's shorelines and natural systems through the lens of climate change, social equity and environmental justice. Nine proposed solutions were revealed after a year-long collaborative process with one specifically entitled "South Bay Sponge" – formulated by a team led by landscape architect, James Corner, and his firm, Field Operations. Their solution deployed adaptive strategies and sponge infrastructure tactics for an area encompassing twenty miles of shoreline in six cities around the original Silicon Valley (RBD website). In *National Geographic*'s April 2019 Special Issue on Cities, the sponge enters the mainstream when SOM (acronym for the architecture firm's founding partners, Skidmore, Owings & Merrill) cites the sponge city as a guiding principle in their proposal for future cities (*National Geographic*, p. 23).

Berlin's updated 2017 plan, northern California's RBD and SOM's future cities concept effectively illustrate China's influence on 21st century international resilient design efforts and the climate crisis. The sponge concept has also reinvigorated the international lexicon on ecological design and green urbanism, especially as an adaptive mechanism for dealing with the climate crisis, the world's *zeitgeist*. China's SCPP emerged in the swift, pro-active and seemingly effective top-down approach with Xi's 2013 speech, China's FYP process and multi-agency effort that led to the selection of thirty pilot sponge cities by 2016. The SCPP blue (water) and green (vegetation) experiment and pending assessment process, and the coming related plethora of published reports, will soon shed light on its effectiveness. China's SCPP also demonstrates their growing international reputation as a leader for environmental stewardship, and effective deployment of rapid-fire media communications and information technology in the 21st century space of flows. China's urgency for sponge cities, low carbon cities, forest cities and various green urban and rural initiatives are driven by dramatic devastation to their natural resources caused by decades of unregulated growth, concerns for responsible nation-building and public health, and the global common good of creating a sustainable future.

Revolutionary praxis, trends and Beautiful China

A glance at China's revolutionary praxis sheds light on the emergence of China's 21st century green revolution and ways environmentalism, greening and the making of purpose-built parks contributed to nation-building. It also speculates on China's emergent ecological modernization, trends in greening and park-making, and the vitality of the maturing discipline of landscape architecture to carry forward Xi's Beautiful China. Soon after Xi was elected by the National People's Congress in March 2012, China experienced extreme summer flood events with similar extreme events the following 2013 summer since Xi took the helm. In this context of ongoing extreme weather events, Xi has framed his legacy as a world leader battling climate change and restoring that nation's environmental image as Beautiful China.

Twenty-first century China has surpassed and transcended the cultural and political narrative surrounding the century of humiliation – the period of colonial modernity between the mid-19th century Opium Wars and the establishment of the People's Republic of China in 1949 after Mao's communist revolution – their second 20th century revolution. China's first revolution occurred in 1911 when Sun Yat-sen's leadership toppled centuries of imperial rule over a feudal society. Public parks and greening during China's first three decades of the 20th century symbolized modernity and freedom from feudal society and imperial tyranny, and the hope for a modern unified democratic nation, especially following Sun's death in 1925. It also overlapped with public gardens for foreign consumption in the treaty port cities. As noted in an earlier chapter, the anniversary of Sun's death (12th of March, currently known as National Arbor Day) in China is marked by voluntary tree planting festivities throughout the nation.

Mao's communist China experienced Soviet-style outdoor recreational spaces as utilitarian places where the masses restored themselves as productive workers to build a socialist society. In Mao's later pronounced period of scarcity, chaos and contradiction, urban public parks were generally converted for productive (agriculture, fisheries, etc) purposes and the rural peasant landscape became a place for agricultural experimentation and destination where city dwellers were sent for re-education. In the dark and violent days of Mao's Cultural Revolution (China's third 20th century revolution), public spaces in Chinese cities became temporary places to "destroy the four olds" and humiliate individuals, especially the urban elite or bourgeoisie who represented the "four olds".

China's fourth (economic) revolution is attributed to Deng's reforms that opened China to the world and were formalized through institutional changes adopted in their 1982 Constitution. Prior to these constitutional changes and after Deng's rise to power, the Stars group of artists exhibited their work unofficially on fences at a Beijing public park adjacent to the National Art Gallery in late 1979. Their unofficial exhibition symbolized freedom of expression and cultural awakening from the dark Mao era, only to be taken down a day later – causing an immediate outcry by the artists and a promptly held public march and demonstration in the streets of Beijing. Public parks and public spaces in Beijing's early post-Mao era were places for creative expression and battles with the central government for artistic individual expression. Around the same time in 1979, hints of urban greening and park-making emerged as the four 1979 inaugural Special Economic Zones (SEZs) grew in southern China. However, landscape architecture would emerge in a more meaningful way after 1982. In other words, while cultural awakening was taking place in the arts and literature in China's late 1970s, the discipline of landscape architecture re-awakened following the adoption of the 1982 Law on the Protection of Cultural Relics, and the 1982 announcement of the first batch of forty-four National Scenic Areas by the then Ministry of Construction (replaced in 2008 with the Ministry of Housing and Rural-Urban Development), a critical national-level agency with responsibilities for housing development and regulating construction nationwide.

China's 1982–1991 "start-up" period for landscape architecture praxis involved work on renovations of park trails and new development of functional trail path networks in National Scenic Areas. Additionally, new Chinese Picturesque parks were built at cultural and historical urban sites as landscape architecture praxis re-awakened in local design institutes, city bureaus and universities. As China grew from 190+ cities in 1978 to 290+ cities in 1984 and 350+ cities in 1990, landscape architects were part of "urban fever" and cosmetic city-making movements with public parks and squares representative of the western neo-classical and Chinese Picturesque design traditions (Yu & Padua 2007; Padua 2014). In this post-Mao start-up period, the discipline was in its embryonic stages and undergoing a mimetic period of re-visiting past traditions and coping with inherited Soviet models. At the same time, some students were sent to study abroad. The ten-year hiatus of non-functioning educational institutions during China's Cultural Revolution was problematic for learning contemporary international landscape architecture. In terms of "meaning", purpose-built projects through the 1990s in China were largely for beautification and functional purposes, particularly in the case of trail systems for National Scenic Areas, with little, if any design innovation. This trend further intensified following the 1990 City Planning Law which triggered formulaic spatial requirements for urban greening and the making of public parks and squares.

By the mid to late 1990s, uncontrolled urbanization devastated China's natural resources, displaced natural systems and polluted the environment. At the same time, Deng's reforms and central government's fiscal devolution fostered tremendous local competition and an entrepreneurial milieu for local government leaders, and more critically the rise of secondary cities in China. Fully opened to the world, foreigners engaged in the planning and design of China's outdoor environments and native-born Chinese returned from overseas studies to lead design institutes and landscape architecture programs in the myriad of China's academic institutions. Simultaneously, access to the worldwide web and Chinese translation of numerous western publications emerged in everyday design practice. In this context, the fusion of international design influences, Chinese Picturesque design genre, and mayors' ambitions to remake their city's identity for a new post-Mao society gave rise to design innovation in China's secondary cities with the cultural production of hybrid modern public parks.

In some ways, the book's case study analysis of the four hybrid modern projects indicate these represent the vanguard projects for the discipline of landscape architecture and contributed to the development of a new urban aesthetic during China's period of hybrid modernization that overlapped the 20th and 21st centuries. Each of the four case study projects celebrated China's post-Mao urban condition and local cultural heritage, and demonstrated ways that landscape architecture in China could mitigate environmental harm created by unregulated growth. The book's four breakthrough projects were required destinations for government officials to visit and contributed to the maturation of landscape architecture praxis in the early years of the 21st century.

The green revolution, the first revolution in China's 21st century, emerged with Hu's 2007 proclamations for developing China as a harmonious society and ecological civilization as the nation prepared for the global spectacle of the 2008 Beijing Olympics and 2010 Shanghai World Expo. Sasaki & Associates' master plan for the Beijing Olympic Green (selected through an international design competition) was largely unimaginative, especially in comparison to the "pomp and circumstance" of the Beijing Olympics opening ceremony curated by Zhang Yimou, internationally known film director. Zhang's creative portrayal reimagined China as an age-old world civilization and dramatically celebrated their 21st century arrival as a major "actor" on the world stage. It's important to note that local landscape architects were engaged to implement Sasaki's concept. The monumental scale of Beijing's Olympic Green made it difficult for the author to "read" the "poetry and pragmatism" Sasaki intended for their interpretation of China's myths and legends combined with contemporary sustainable development during a visit in 2008. Most cities in China were engaged in park-building and sprucing up their cities in anticipation of spill-over tourism from the Beijing Olympics. These greening and park-making trends, combined with the "spectacle" designed landscapes of the 2008 Beijing Olympics and 2010 Shanghai World Expo, illustrate a nation-building period in China's first decade of the 21st century when outdoor designed environments were created for international consumption (Figure 6.4). A different experience was underway in western China near Chengdu, Sichuan Province where public open spaces in and around the epicenter of the May 2008 "Great Wenchuan earthquake, *Wenchuan da dizhen*", were being deployed as temporary shelter for housing local residents and staging grounds for disaster relief.

In alignment with Hu's notion for the scientific development of a harmonious society and China's 11th FYP, the State Council launched the 2008 Urban and Rural Planning Law (URPL), *Chengxiang Guihua Fa*. It replaced the 1990 City Planning Law with an expanded scope to include oversight of China's urban-rural spatial distribution and integration of cities, towns, townships and villages, and influenced the discipline of landscape architecture. It included a mandated thirty-day period of public review intended to solicit feedback from experts and the public before the city or towns' proposed or revised twenty-year plan document was submitted to higher administrative authorities for approval. In political terms, this public review was considered significant with the central government sending a signal to prevent "image" or "achievement" projects and random amendments to previously approved master plans; and it introduced a democratic process that enabled communities to provide input on the details of development in their cities and towns (Zeldin 2007). The 2008 URPL also invoked an environmental imperative where the logic of land allocation was based on the conservation of natural resources and goals to reduce pollution, improve the environment, promote energy efficiency, and protect land suitable for cultivation and natural resources (Curien 2014).

As cities in China were dealing with the new URPL, and the nation and world were preparing for the post-Olympics spectacle of the 2010 Shanghai World Expo, 2008 brought with it the global financial crisis now known as the "Great Recession". To ward off national economic concerns, and to an extent international concerns, Hu proactively launched a major economic stimulus package valued at 4 trillion RMB (US$586 billion) in early November. While some of the stimulus funds had already been allocated for the Sichuan post-earthquake reconstruction, the distribution was

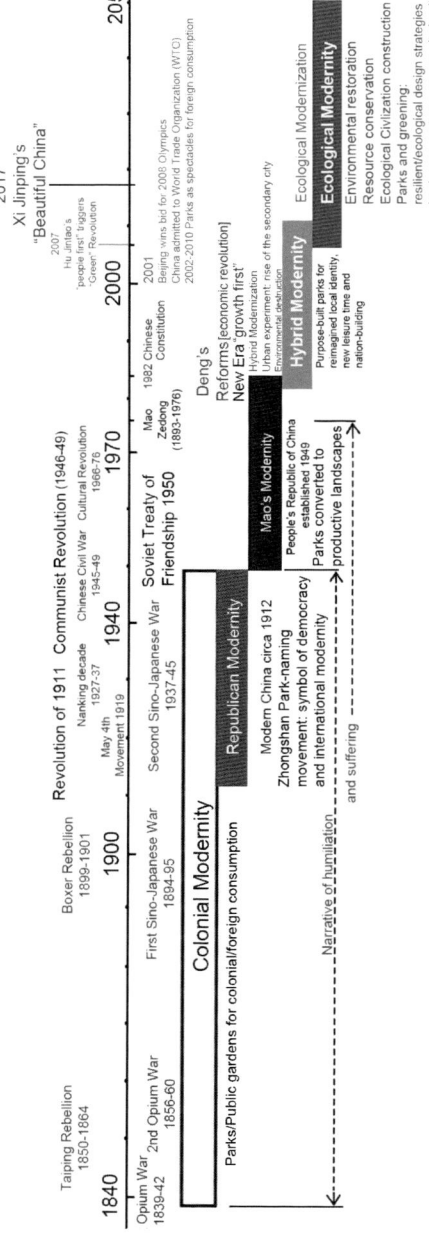

Figure 6.4 Timeline of China's park history, themes and revolutionary praxis, graphic by author

proposed to take place by the end of 2008, and completely allocated by February 2009 (Naughton 2009). These funds were allocated to certain sectors: urban and rural public infrastructure development; and environmental investment, pollution control and protection of natural resources (Naughton 2009). The stimulus package enabled China's ongoing "green" experimentation and opportunities for the maturation of landscape architecture.

In the outer fringes of urban Shanghai, the 2010 World Expo featured Turenscape's 2009 completed work on Houtan Park: the conversion of a brownfield site into a blue-green "living machine" that cleansed polluted water through biomimetic processes. Houtan Park's design concept was similar to Chengdu's Living Water Park, the book's first case study project in Sichuan Province, completed eleven years before and designed by Betsy Damon and Margie Ruddick. On an island near Shanghai, the first phase of Dongtan, the "world's first eco-city" by Arup Associates, a London-based multi-disciplinary firm, failed to materialize due to corruption when the Shanghai 2010 World Expo opened its doors (Larson 2009). Other eco-cities on greenfield sites, designed by foreign experts on sustainable design, also failed due to financial challenges and lack of construction experience. However, China's emergent "green" urban experiment gave rise to successful projects, even with failed eco-cities designed for global consumption. For example, across town from the 2010 Shanghai World Expo site is the culture-based, environmentally oriented community project known as Gubei Promenade, a project designed by the SWA Group (Los Angeles and Shanghai branch offices), landscape architecture and planning practice.

SWA's pedestrian-only Gubei Promenade demonstrated to the surrounding community and Shanghai the importance of urban tree canopy for carbon sequestration, creating new natural and wildlife habitats, as well as promoting healthy communities (Padua 2012, 14). It's also important to note that Shanghai emerged as a creative hub where many award-winning national and international landscape architecture and interdisciplinary firms established offices and maintained their presence. For example, in addition to the SWA Group, foreign-owned firms like EDSA, AECOM, Design Land Collaborative and GVL-Gossamer have offices in Shanghai with goals to work on projects in China and other parts of Asia. Turenscape also has a branch office in Shanghai, along with a new generation of award-winning landscape architects including Z + T Studio, a firm founded by native-born Zhang Dong and Tang Diying, both educated at the University of Massachusetts with professional experience at Martha Schwartz's Boston office. Shanghai is also home to Tongji University, one of the oldest institutions in China to offer classes on landscape studies with influences from Bauhaus modernism and the Chinese Picturesque, as espoused by Professor Chen Congzhou (1918–2000), noted landscape theorist and a leading historian of his generation. According to I. M. Pei (2009 interview), Chen designed the Chinese Picturesque garden at Pei's 1982 Fragrant Hill Hotel project in Beijing.

As Hu was advocating for China's harmonious society and green economic reforms, the State Council approved the professional master's degree in landscape architecture in 2005 for delivery in twenty-five universities (Li 2017). As the 11th FYP top-down policy directives made its way to local government, the shift towards environmentalism and resource efficiency began driving China's urban-rural nation-building experiment. Landscape architects participated in the making of low carbon cities and healthy cities, with the deployment of blue-green infrastructure (sponge) and ecological design strategies in parks and open space systems that restored natural systems and created new urban wildlife habitats, protected natural resources, mitigated impaired environments, and battled climate change. This work was also part of the next stage of China's vast nation-building effort and major shift to ecological modernization, especially in light of the 11th FYP policy directive for a circular economy, resource efficiency and environmental restoration. As the second decade of the 21st century commenced, China's State Council announcement on modifications to discipline rankings called for a major celebration among landscape architects. This 2011 official ranking gave parity to "Landscape Architecture" as a so-called national "First Level" discipline along with Architecture and Urban-Rural Planning. In China, these three First Level national disciplines fall under the national key discipline category of "Engineering".

Xi Jinping was elected China's President in March 2012 at the same time the concept of "ecological civilization construction" was formalized in their Constitution. In this policy context, ecological civilization was placed on par with "economic, political, cultural and social progress" (Batchell 2018). Within months of Xi's election, China experienced extreme summer flood events and Xi soon advocated for the development of sponge cities, a national-level policy initiative. As noted earlier, China's top-down policy-making, authoritarian government, and action-oriented institutional apparatus enabled the launch of the SCPP within months of Xi's election. Yu Kongjian, an international award-winning landscape architect based in Beijing and Turenscape founder, is a staunch CPC member who has been instrumental in China's SCPP. Turenscape's Yanweizhou Park in Jinhua, Zhejiang Province serves as an official exemplar for China's SCPP (Padua & Lung 2019). Yu also serves as Vice Chair (one of three) on the National Sponge City Technology Committee comprised of thirty-three members from various ministries and experts from various fields (Wu & Padua 2019).

The scope and scale of projects by landscape architects working in China has increased tremendously and can be attributed to the central government's acknowledgment of the discipline's significance. China's complex top-down governance involves a web of official activities including announcements by the president at official meetings, related political program campaigns and a flurry of FYP-related initiatives launched by various ministries and central government administrative units. This flurry of initiatives appears ad hoc and chaotic as China scrambles to meet Xi's pledge for a "Beautiful China" announced at the 19th National CPC Congress (held every five years) in October 2017. In this light, Xi's Beautiful China builds on the ancient Daoist philosophy of "unity with nature" with an emphasis on a renewed effort to mitigate environmental devastation caused by decades of unrestricted growth. Xi noted the importance for the CPC to "meet people's ever-growing demands for a beautiful environment" or a clean environment is critical for greater happiness and well-being; and he asserted the importance of China to take on the leadership role as a formidable global environmental steward (Xi 2019). The opportunities for the discipline of landscape architecture are linked to reclaiming the importance of China's natural resources in Xi's interpretation of Beautiful China.

Much of the work carried out in the discipline of landscape architecture intersects with Xi's vision of Beautiful China and align with directives in the 13th FYP 2016–2020, especially infrastructure and water safety, the new type of urbanization for livable cities, and environmental restoration and protection. Currently, the discipline of landscape architecture in China is multi-faceted and sometimes involves work with other disciplines. Some projects are investigated through research institutes with central government financing. Local projects are financed by local governments and private entities (domestic or international). Some examples of national initiatives launched in 2017 include a new system of national parks and the establishment of four large conservation-oriented parks; the "pastoral complex, *tianyuan zonheti*", rural revitalization strategy; and the "Guidance on Promoting Green Belt and Road", essentially 2017 changes to the 2013 mega-infrastructure Belt and Road initiative. Central government subsidies for China's thirty pilot sponge cities end in 2020 and the State Council set targets for retrofitting Chinese cities with the re-use of up to 80% of rainwater by 2030 (Wu & Padua 2019). This implies work for landscape architects will be sustained. However, while the discipline of landscape architecture generally blends art and science, it appears that this blend is tipping towards science in China.

In the current mode of ecological modernization and the national pursuit of Beautiful China, the discipline of landscape architecture appears influential. The sponge infrastructure efforts and development of four vast national parks contribute to the lifeline of the discipline, as does many of the critical activities needed to clean and restore China's urban and natural environments. However, as a colleague based in south China since the 1990s notes, much of the recent work in China by landscape architects has fallen into the trap of "pattern-making". In some ways this work can be defined as simply "green", resilient and sustainable, and portrays visual monotony or lack of creativity. In this light, finding "poetry" or beauty in China's designed landscapes for design critics and historians may prove challenging in the coming years. For the western world, China's first two decades of the 21st century filled voids for international landscape architects, particularly given the so-called lost American decade marked by the Great Recession and the horrific events in New York, Washington, D.C. and Pennsylvania on 11 September 2001. China's open-door policy along with its tremendous urban experiment and green revolution during the 21st century enabled China's growth of the discipline of landscape architecture in a compressed period of time.

China's top-down political context strongly influences the discipline of landscape architecture and allied professionals working in the built and natural environments, and perhaps even more with the 2018 Constitution changes that eliminated term limits for president, effectively making Xi president for life. While "green" is the current basic aesthetic in a nation-building mode of ecological modernization, innovation in China's designed landscapes may become hindered given Xi's turn to neo-traditionalism – a type of Confucian narrative for traditional culture. Xi revealed this concept to artists and writers during a November 2016 speech when he called on them to devote their efforts to "propagating traditional culture" and "extract essence and draw energy from the treasure vault of Chinese culture" (Klimes & Marinelli 2018). Soon after Xi's speech, the State Council released the official document entitled "Opinions on the Implementation of the Development of Outstanding Traditional Chinese Culture" in January 2017, and is antithetical to Mao's disdain for "old culture". The following excerpt from this official document hints at China's near-term physical development:

> Dig deep into the historical and cultural values of the city, define a selection of classical elements and iconic symbols that highlight cultural characteristics, integrate urban-rural construction, urban planning and design, and reasonably apply them to public spaces such as urban sculptures, squares and gardens to avoid monotonous and thousand cities.
>
> (Xinhua 2016)

Combine this excerpt with qualitative language from the earlier 2016 "Guidelines for Green and Smart Urban Development" – forbidding bizarre architecture and advocating for the preservation of distinctive landscape and cultural identities of cities, and a seemingly uncreative schema emerges for parks, landscape systems or outdoor designed environments and the remaking of cities in China. This philosophical move in the direction of traditional culture suggests cultural conservativism and could mean a regression to the Chinese Picturesque for the discipline of landscape architecture.

Xi's Beautiful China is imbedded with an imperative for a clean environment by 2050 that is simultaneous with cultural conservativism and a return to Confucian classics. Trends in the discipline of landscape architecture may change some in the next thirty years and will most likely be linked to the "cleaning" or restoration of China's environment. A debate among world scientists continues in terms of the degree that environmental degradation could be arrested or restored in China, given the vast disruption to natural systems and pollution caused by hyper-urbanization. However, numerous national sustainability programs and related interventions that commenced in 1998 indicate significant positive impacts, and are expected to improve when combined with the specificity of China's 13th FYP directive for nationwide spatial zoning with specific areas designated for agricultural production, forest conservation and development, urban development, and key ecological functions. Adding to these efforts is China's establishment of four new national parks, vast in scale, with three primarily for habitat conservation (snow leopard, giant panda and Siberian tiger and Amur leopard) and the fourth protecting the sources of three major rivers (Yellow, Yangtze and Mekong). In this context, the discipline of landscape architecture will continue to engage in a range of efforts from land planning and conservation to detailed low impact site design. Additionally, in terms of ecological modernization, work by landscape architects on strategic greening through greenways, green belts and integrated systems of open space will contribute to carbon sequestration that improves air quality, as well as soil erosion. Blue-green infrastructure (or sponge tactics) will continue to be deployed and improve water quality and urban stormwater management, attenuate extreme flood events, as well as bolster water conservation.

Defining "poetry" in China's near future designed landscapes may need to be reframed, given the limits of a green or naturalistic aesthetic in the contemporary state of ecological modernization. For those in the discipline of landscape architecture who defy the Chinese Picturesque, a design vocabulary that could dominate outdoor designed environments given the call for traditional culture, design innovation ostensibly could emerge in the realm of landscape materials and tectonics as part of the circular economy (net-zero waste or resource efficiency through recycling and re-using). For example, a closer reading of the Sponge City Technical manual reveals specifications for construction materials not made in China (Wu & Padua 2019). An exploration of recycled materials could prove useful, for example, in topographic landform manipulation (landscape tectonics) and the embodied energy in "hard" (vs soft vegetation) materials critical for sponge city implementation. "Big data" and emergent best management practices for the life cycle of materials, carbon sequestration and the shifting green paradigm may also factor into this dimension and the expanding role of landscape architecture.

At the same time Xi is pushing for a clean environment, "Made in China 2025", a program launched in 2015 could offer opportunities for materials innovation needed for the effectiveness of sponge cities, smart cities and the like. Chinese policy makers were inspired by Germany's Industry 4.0 Development Plan and Made in China 2025 deploys a similar strategy to update their manufacturing base through rapid development of ten high-tech industries. The discipline of landscape architecture and allied disciplines working in the natural and built environments can readily intersect with Made in China 2025 activities, especially in the "green" and "smart" development of places where advanced robotics, artificial intelligence (AI), next generation information technology, and electric and alternative energy vehicles are tested and utilized. In fact, instructors and students in landscape architecture and urban design studio courses around the world have been exploring "driverless cities" and the adaptive re-use of remnant spaces left over from unnecessary road infrastructure, as well as related changes in social behavior.

Meeting Xi's goal for Beautiful China by 2050 is a challenge in itself, especially given the metrics, or lack thereof, for environmental restoration and debate surrounding irreparable damage to China's natural resources. However, efforts by the discipline of landscape architecture could lead thinking for a reimagined "green" emergent mid-21st century aesthetic in light of the climate crisis and human-centered development. China's ongoing urban-rural experiment creates a crucible where the aesthetics of "new" or "second" and potentially "third" nature in China's context can be explored by landscape architects, especially within the contemporary context of ecological modernization. China's new generation of landscape architects have the opportunity to build from the radical notion for "second nature" that reconciles urban and rural contexts to the benefit of both (Geuze & Skjonsberg 2010). In some ways, Turenscape's "designed ecologies" initiates thinking along this trajectory (Saunders 2012). However, with the thirty-year time horizon for achieving Beautiful China, temporality enables deeper theoretical and practical investigations by the next generation(s) of landscape architects and allied disciplines working in China's urban and rural contexts. This temporality elucidates tremendous hope for the discipline of landscape architecture and its ongoing evolution in China and the world.

References

Barboza, D. (2006) I. M. Pei in China, Revisiting Roots, *New York Times*, 9 October.

Batchell, J. (2018) China Builds an 'Ecological Civilization' While the World Burns, *People's World*, peoplesworld.org, 21 August.

Berlin Journal. (2017) Berlin Plans to Implement Climate Change Mitigation Strategies, *Berlin Journal*, 23 August.

Cheung, T. & He, H. (2019) China's State Council Reveals Details of Greater Bay Area Plan to Turn Hong Kong and 10 Neighboring Cities Into Economic Hub, *South China Morning Post*, 19 February.

Curien, R. (2014) Chinese Urban Planning Environmentalising a Hyper-Functionalist Machine? *Chinese Perspectives*, 3(1), 23–31.

Economy, E. & Zhang, X. (2015) A Small Chinese Experiment With Large Environmental Implications, *Diplomat*, 28 August.

Fan, C. (2006) China Eleventh Five-Year Plan (2006–2010): From "Getting Rich First" to "Common Prosperity", *Eurasian Geography and Economics*, 47(6), 708–723.

Fang, C. (2015) Important Progress and Future Direction of Studies on China's Urban Agglomerations, *Journal of Geographical Sciences*, 25(8), 1003–1024.

Geuze, A. & Skjonsberg, M. (2010) Second Nature New Territories for the Exiled. In: *Landscape Infrastructure Case Studies by, SWA*, Hung, Y. & Aquino, G., Basel, Switzerland, Birkhauser.

Giddens, A. (1998) *The Third Way*, Cambridge, Polity Press.

Gipouloux, F. & Li, S. (2015) The City Creation Process in China. In: *China's Urban Century: Governance, Environment and Socio-Economic Imperatives*, Gipouloux, F. (ed.). Northhampton, MA, Edward Elgar Publishing Ltd.

Harvey, D. (1996) *Justice, Nature and the Geography of Difference*, Oxford, Blackwell.

Heilmann, S. & Melton, O. (2013) The Reinvention of Development Planning in China, 1993–2012, *Modern China*, 39(6), 580–628.

Ivy, R. (2007) Suzhou Museum' by I. M. Pei, *Architectural Record*, May.

Jia, H., Wang, Z., Zhen, X., Clar, M. & Shaw, L.Y. (2017) China's Sponge City Construction: A Discussion on Technical Approaches, *Frontiers of Environmental Science and Engineering*, 11(18).

Klimes, O. & Marinelli, M. (2018) Introduction: Ideology, Propaganda, and Political Discourse in the Xi Jinping Era, *Journal of Chinese Political Science*, 23(3), 313–322.

Kuo, L. (2019) Chengdu is Blossoming as China's 'Park City', But Its Residents Pay the Price of Beautification, *South China Morning Post*, 7 February.

Larson, C. (2009) China's Grand Plans for Eco-Cities Now lie Abandoned, *Yale Environment*, 360, 6 April, Yale University.

Leach, A. (2016) Soak It Up: China's Ambitious Plan to Solve Urban Flooding with 'Sponge Cities', *The Guardian*, 3 October.

Li, X. (2017) *Graduate Education in Landscape Architecture*, Conference of Educators in Landscape Architecture (CELA)/Chinese Landscape Architecture Education Committee (CLAEC) Conference, 26–29 May 2017, Beijing.

Li, Y. (2019) Suzhou Industrial Park Celebrates 25 Years, *China Daily*, April 12.

Li, Q., Wumaier, K. & Ishikawa, M. (2019) The Spatial Analysis and Sustainability of Rural Cultural Landscapes: Linpan Settlements in China's Chengdu Plain, *Sustainability*, 11(16), 4431.

Ma, C. (2015) Why Did China Convene its First Urban Work Conference in 37 Years? *China Daily*, 30 December.

Mol, A. (1992) Sociology, Environment, and Modernity: Ecological Modernization as a Theory of Social Change, *Society and Natural Resources*, 5(4), 323–344.

Naughton, B. (2009) Understanding the Chinese Stimulus Package, *China Leadership Monitor*, 28, 1–12.

Padua, M. (2003) Industrial Strength, *Landscape Architecture*, 93(6), 76–85 and 105–107.

Padua, M. (2004a) Future Scale, *Landscape Architecture*, 94(8), 106–115.

Padua, M. (2004b) Teaching the River, *Landscape Architecture*, 94(3), 100–109.

Padua, M. (2012) This Way Shanghai, *Landscape Architecture*, 102(12), 54–65.

Padua, M. (2014) China: New Cultures and Changing Urban Cultures. In: *New Cultural Landscapes*, Roe, M. & Taylor, K., (eds.). London, Routledge.

Padua, M. & Lung, S. (2019) Adaptive Landscapes for Coastal Restoration and Resilience in Contemporary China. In: *Sustainable Coastal Design and Planning*, Mossop, E., (ed.). Boca Raton, Florida, CRC Press.

Saunders, W., (ed.) (2012) *Designed Ecologies: The Landscape Architecture of Kongjian Yu*. Basel, Switzerland, Birkhauser.

Shi, S., Kondolf, G. & Li, D. (2018) Urban River Transformation and the Landscape Garden City Movement in China, *Sustainability*, 10(11), 1–20.

SOM (2019) Proposal for Future Cities, *National Geographic*, April.

Wang, Q. & Guo, R. (2010) Threat of Worst Flood in 12 years, *China Daily* 15 July.

Wang, K. & Liu, J. (2017) The Spatiotemporal Trend of City Parks in Mainland China Between 1981 and 2014: Implications for the Promotion of Leisure Time Physical Activity and Planning. *International Journal of Environmental Research and Public Health*, 14(10), 1150–1160.

Wang, Y., Song, Q., He, J. & Ye, Q. (2015) Developing Low-Carbon Cities Through Pilots, *Climate Policy*, 15(1), 81–103.

Watts, J. (2015) China: The Air Pollution Capital of the World, *The Lancet Journal*, 366(9499), 1761–1762.

Workman, J. (2017) Sponge Cities: Can China's Model go Global? *Source*, August, International Water Association.

Wu, J. & Padua, M. (2019) *Investigating China's Sponge City Pilot program*, CELA Annual conference, 6–9 March 2019, California, University of California Davis.

Xi, J. (2019) Pushing China's Development of an Ecological Civilization to a New Stage, *QiuShi Journal (English edition)*, 11(2), 39.

Xinhua (2014) City Classifications New-Type Urbanization (2014–2020), *Xinhua*, 16 March.

Xinhua (2016) China Outlines Roadmap to Build Better Cities, *Xinhua* 21 February.

Yu, K. & Padua, M. (2007) China's Cosmetic Cities: Urban Fever and Superficiality, *Landscape Research*, 32(2), 225–249.

Zeldin, W. (2007) China: New Urban and Rural Planning Law, *Global Legal Monitor*, 2 November.

Zhao, R. (2013) Solving the Problem of Urban River Pollution: Protect the River from the Headwater and Restore the Ecosystem, *NGO Case Study Series*, Funded by Ford Foundation.

Zheng, J. (2016) China Looks to Regulate City Growth, *China Daily*, 22 February.

Zhou, N., He, G. & Williams, C. (2012) *China's Development of Low Carbon Eco-Cities and Associated Indicator Systems*. LBNL-5873E, Berkeley, CA, Lawrence Berkeley National Laboratory.

Index

Note: Page numbers in 'italics' refer to figures 'bold' refer to tables.